Mathematics

Essential Skills

3

Level 8, 737 Bourke Street, Docklands, Victoria 3008, Australia

Oxford University Press is a department of the University of Oxford. It furthers the University's objective of excellence in research, scholarship, and education by publishing worldwide. Oxford is a registered trademark of Oxford University Press in the UK and in certain other countries.

© Pat Lilburn 2011

First published 2011
Reprinted 2014 (twice), 2023

Reproduction and communication for educational purposes
The Australian *Copyright Act 1968* (the Act) allows a maximum of one chapter or 10% of the pages of this work, whichever is the greater, to be reproduced and/or communicated by any educational institution for its educational purposes provided that the educational institution (or the body that administers it) has given a remuneration notice to Copyright Agency Limited (CAL) under the Act.

For details of the CAL licence for educational institutions contact:

Copyright Agency Limited
Level 15, 233 Castlereagh Street
Sydney NSW 2000
Telephone: (02) 9394 7600
Facsimile: (02) 9394 7601
E-mail: info@copyright.com.au

Reproduction and communication for other purposes
Except as permitted under the Act (for example, any fair dealing for the purposes of study, research, criticism or review) no part of this book may be reproduced, stored in a retrieval system, communicated or transmitted in any form or by any means without prior written permission. All enquiries should be made to the publisher at the address above.

ISBN 978 0 19 557401 2

Typeset by Palmer Higgs
Printed in China by Golden Cup Printing Co. Ltd

PAPUA NEW GUINEA

Mathematics

Essential Skills

3

Pat Lilburn

OXFORD
UNIVERSITY PRESS
AUSTRALIA & NEW ZEALAND

Contents

For Students

The *Oxford Essential Skills Book* has been written to support the *Oxford Grade 3 Mathematics Student Books A* and *B*.

The focus of this book is on providing practice examples to revise and support the concepts learned in the two Student Books.

The book is set out under the five Strands outlined in the *Lower Primary Mathematics Syllabus* and *Teacher Guide*:

- Number and Application
- Measurement
- Space and Shape
- Chance and Data
- Patterns

Within each Strand, you will find many practice examples that relate directly to the Learning Outcomes for that Strand. These Learning Outcomes are specified in the *Lower Primary Mathematics Teacher Guide*.

The subheading in each Strand shows you what the examples will help you practise.

3.1.1 Count, order, read and represent two and three-digit numbers

The Help Box contains worked examples to show you how to set out and complete the examples in each Strand. The Remember Box contains information that will help you complete the examples.

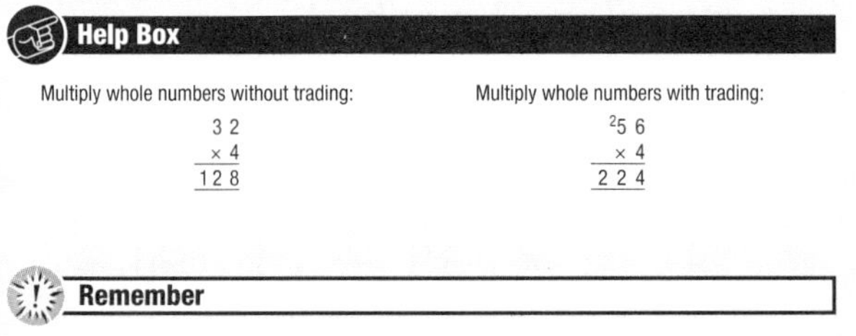

Help Box

Multiply whole numbers without trading:

$$\begin{array}{r} 32 \\ \times\ 4 \\ \hline 128 \end{array}$$

Multiply whole numbers with trading:

$$\begin{array}{r} {}^{2}56 \\ \times\ 4 \\ \hline 224 \end{array}$$

Remember

When the answer to a division number sentence has a remainder, for example 25 ÷ 6, this is how we write it: 25 ÷ 6 = 4 rem. 1

At the end of each Strand there is an Assessment section that covers all the examples from that Strand. If you have difficulty with any test questions, go back to the relevant page in the Strand and look at the worked examples before doing the test again.

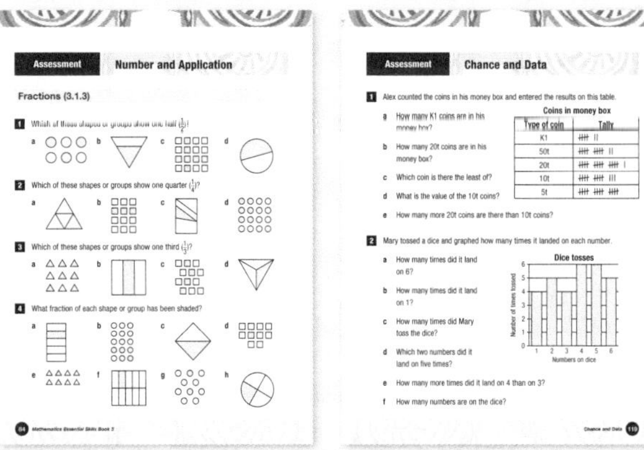

Assessment — Number and Application

Fractions (3.1.3)

Assessment — Chance and Data

There is an Answers section and a section for Important Facts at the back of the book. The Answers section contains answers to all practice examples and test questions while the Important Facts section contains some of the facts that you will need to refer to throughout the year.

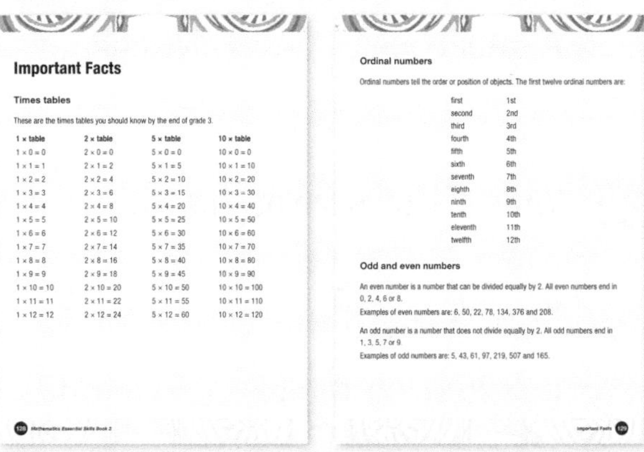

Important Facts

Times tables

Ordinal numbers

Odd and even numbers

Strand Number and Application

Count, order, read and represent two and three-digit numbers

Read and write two-digit numbers in words and numerals

1 Write each number word and its matching number.

a	twenty-seven	207	20	27	**b**	forty-five	450	45	405
c	seventy-one	71	701	170	**d**	eighty-nine	98	89	809
e	thirty-six	36	306	336	**f**	ninety-three	39	93	903
g	sixty-eight	608	68	680	**h**	fifty-four	54	504	45

2 Write each group of number words as numerals and put them in order.

a	fifty-six	twenty-eight	forty	sixty-two	eighteen
b	thirty-one	seventy-three	twenty-nine	ninety-five	twelve

3 Write in words the number that comes **before** each of these numbers.

a	40	**b**	94	**c**	57	**d**	86	**e**	31	**f**	79
g	25	**h**	62	**i**	48	**j**	89	**k**	19	**l**	90

4 Write in numerals the number that comes **after** each of these numbers.

a	fifty	**b**	thirty-seven	**c**	eighty-nine	**d**	twenty-three
e	forty-two	**f**	sixty-one	**g**	seventeen	**h**	ninety-four

Read and write three-digit numbers in words and numerals

1 Write the matching numbers and words.

a	seven hundred and forty-five	361	**b**	nine hundred and seventeen	970
c	one hundred and eighty-four	745	**d**	three hundred and sixty-one	229
e	five hundred and thirty-six	316	**f**	two hundred and twenty-nine	184
g	three hundred and sixteen	917	**h**	nine hundred and seventy	536

2 Write the number that matches each set of words on the left.

a	eight hundred and fifty	850	805	85
b	five hundred and sixteen	165	561	516
c	two hundred and nine	209	290	219
d	seven hundred and thirty-one	713	731	730
e	four hundred and twenty	402	420	204
f	one hundred and ninety-one	190	119	191

3 Write in words the number that comes **before** each of these numbers.

a	160	**b**	106	**c**	200	**d**	139	**e**	462	**f**	720
g	310	**h**	800	**i**	301	**j**	570	**k**	919	**l**	645

Use ordinal numbers

1

a What position is Alice in?

b Who is 4th in the line?

c Who is last in the line?

d Who is 2nd in the line?

e What position is Nina in?

f What position is Esta in?

g Who is 6th in the line?

h What position is Lydia in?

2

a Who will be served 1st?

b What position is Jay in?

c When will Emily be served?

d Who is 11th in the line?

e Who will be served 3rd?

f What position is Mivai in?

g Who will be served straight after the 9th person?

h Who will be served straight before the 7th person?

3 Write the ordinal name of the object that is different in each row. For example, in the first row the 6th object is different.

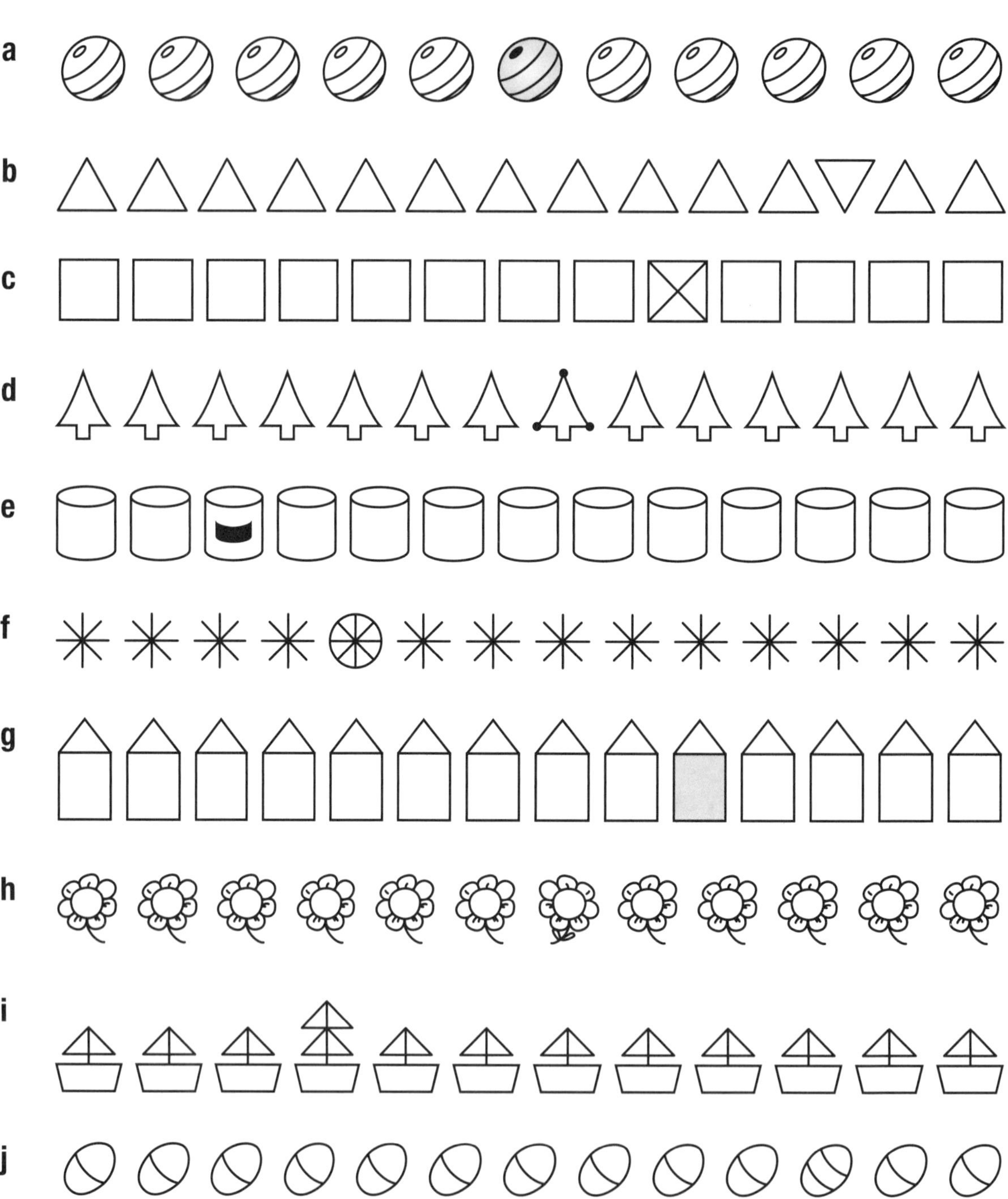

Order two-digit numbers

Write the largest number in each group and then write the smallest number in each group.

1

2 84 63 52 27 56 65

3 51 43 92 36 17 29

4 72 66 58 87 43 30

5 34 42 76 55 49 38

6 29 60 16 68 86 81

7 36 95 58 51 44 89

8 79 24 63 49 27 77

Order two and three-digit numbers

1 Write the largest number in each group, and then write a number larger than it.

a	67	76	16	**b**	43	34	40
c	95	90	59	**d**	68	80	86
e	263	623	362	**f**	508	500	85
g	124	214	421	**h**	517	751	175
i	126	162	216	**j**	463	436	364
k	715	517	571	**l**	608	806	680

2 Write in order three numbers that come between the numbers shown.

a	65	_____	_____	_____	120
b	300	_____	_____	_____	350
c	500	_____	_____	_____	600
d	450	_____	_____	_____	480
e	790	_____	_____	_____	810
f	198	_____	_____	_____	210
g	281	_____	_____	_____	307
h	685	_____	_____	_____	712
i	398	_____	_____	_____	420
j	879	_____	_____	_____	905

3 Read these numbers and then answer the questions below.

	217	604	156
570	182	937	
	411	280	95
722	885	360	
	150	505	67

a Write the numbers that are more than 400. Put these numbers in order from smallest to largest.

b Write the numbers that are less than 200. Put these numbers in order from smallest to largest.

c Write the numbers with a 5 in them. Circle the largest. Cross the smallest.

Represent two and three-digit numbers

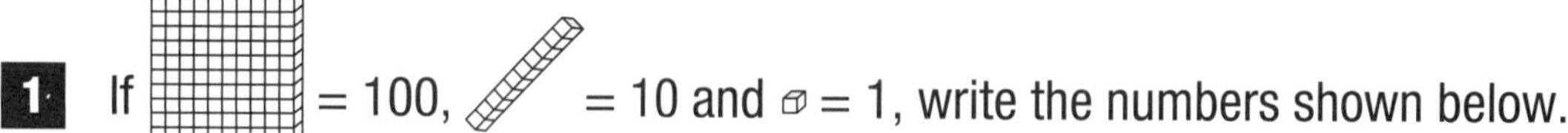

1 If [hundreds block] = 100, [tens rod] = 10 and [unit cube] = 1, write the numbers shown below.

a 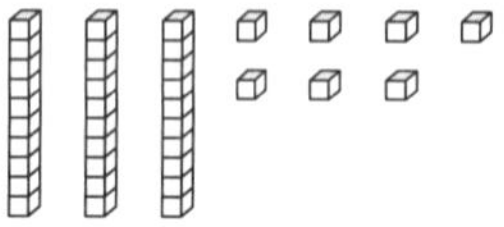	b 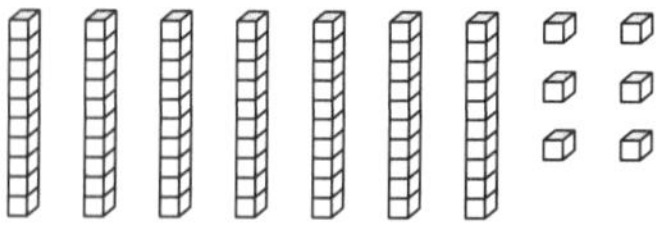
c 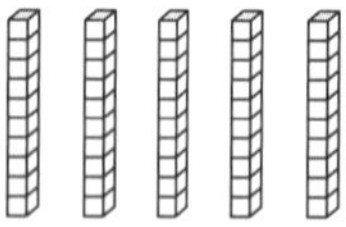	d 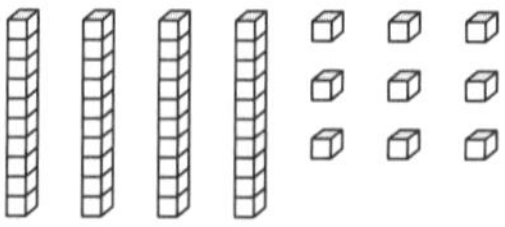
e 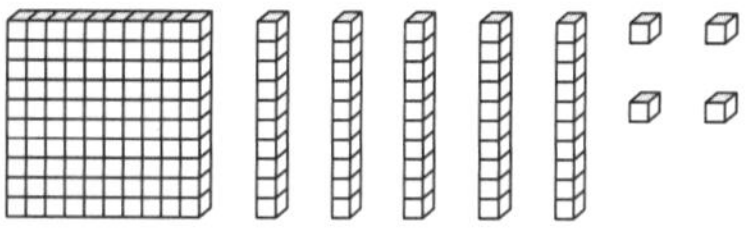	f 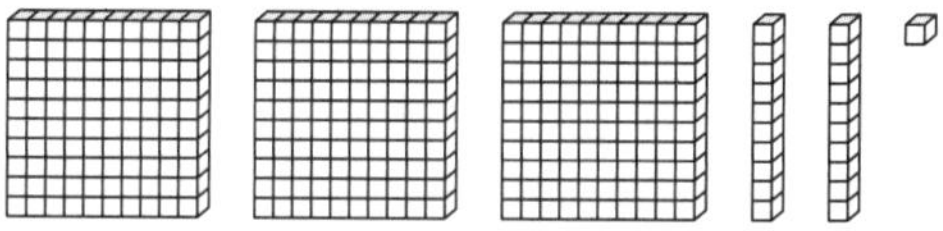
g 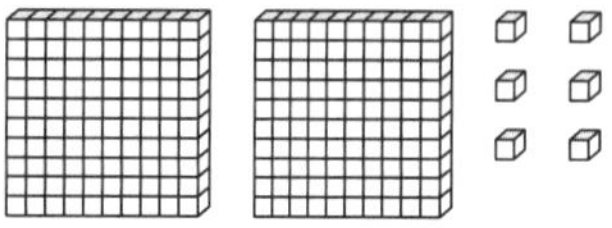	h 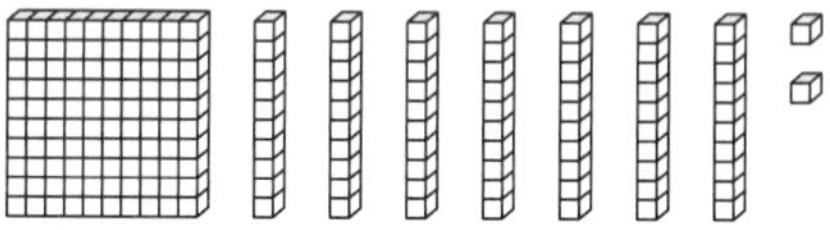
i 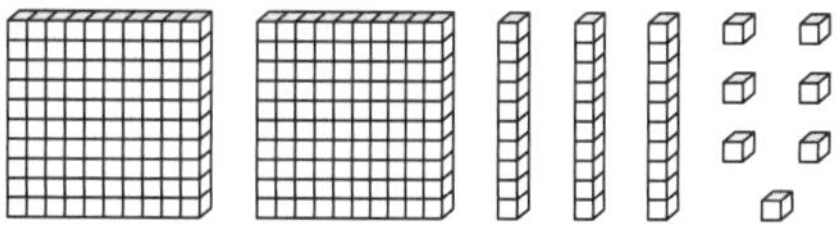	j 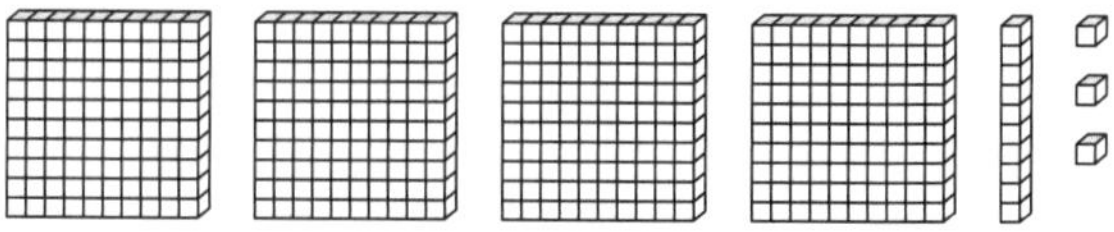

2 What number is shown on each abacus?

a 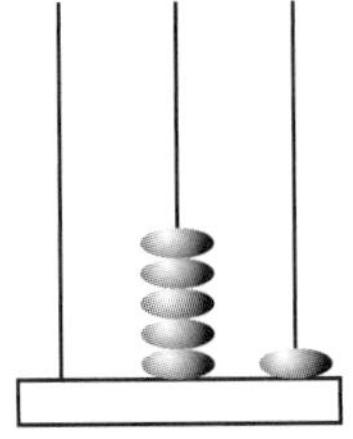

b 

c 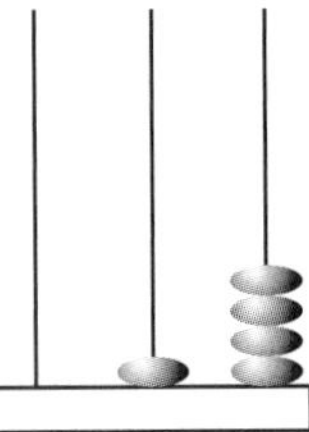

d 

e 

f 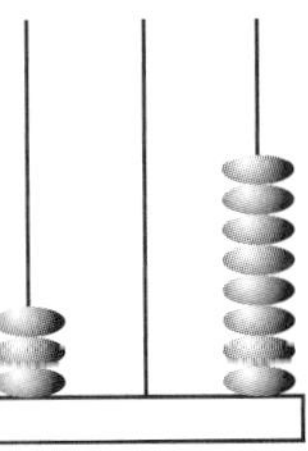

g 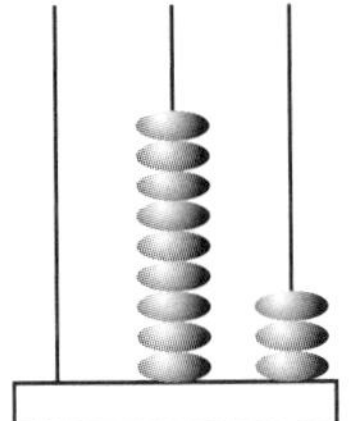

h 

i 

j 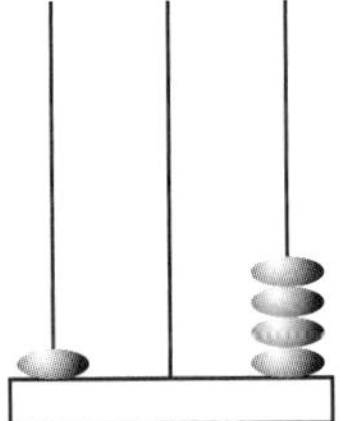

k 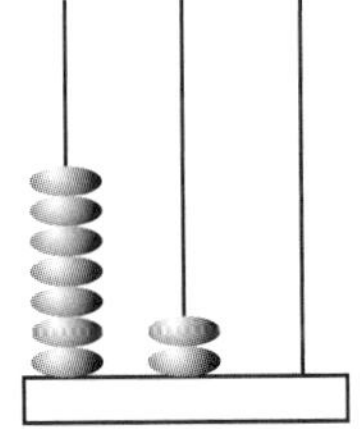

l 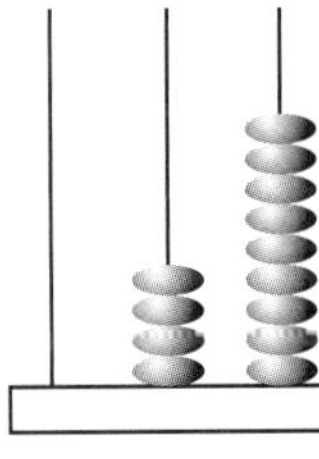

m 

n 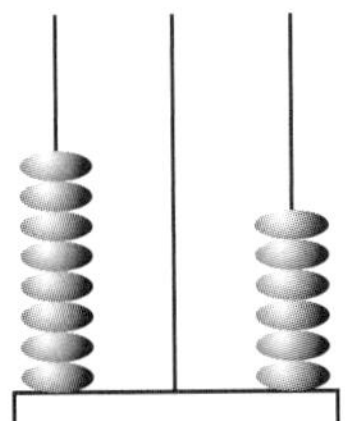

o 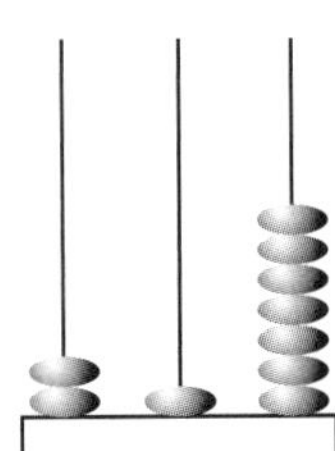

Recognise the value of digits in numbers

1 Copy the picture below and colour it using the following code.

Colour numbers with 5 tens red.	Colour numbers with 3 tens green.
Colour numbers with 8 ones yellow.	Colour numbers with 6 ones blue.

2 Write some more numbers that would be coloured green.

3 Write some numbers that would not be any of these colours.

Make two and three-digit numbers

1 Make as many two-digit numbers as you can from these two digits. [7] [2]

2 Make the largest possible number with each pair of digits.

a [2] [3] **b** [6] [0] **c** [3] [8] **d** [5] [9]

3 Make as many three-digit numbers as you can from these three digits. Use each digit once in each number. [7] [8] [2]

4 Use these digits once in each number to make: [5] [9] [4]

a the largest number

b the smallest number

c the number closest to 500

d the number closest to 900

5 What numbers are these?

a [7] [80]

b [10] [7] [200]

c [40] [9] [500]

d [5] [50]

e [200] [8] [30]

f [7] [10] [300]

g [60] [200] [1]

h [500] [3] [30]

i [700] [10] [2]

j [80] [6] [100]

k [4] [300] [50]

l [70] [9] [400]

m [30] [700]

n [5] [900]

Count by 5s

1 Write the numbers that you will say if you count by 5s starting at 0.

12 20 18 10 8 17
15 3 5 22 35
25 31 38 30 27 40

2 Write the numbers that you will say if you count by 5s starting at 70.

72 80 96 100 78 83
95 85 75 103 110
90 89 105 115 108 92

3 Write the numbers that you will say if you count by 5s starting at 47.

50 54 62 70 52 67
60 57 55 87 75
77 66 58 72 88 82

4 Write the numbers that you will say if you count by 5s starting at 91.

92 101 100 99 87 96
106 103 110 116 121
111 107 114 126 131 94

Count by 10s

1 Count by 10s to find how many sticks. Each bundle contains 10 sticks.

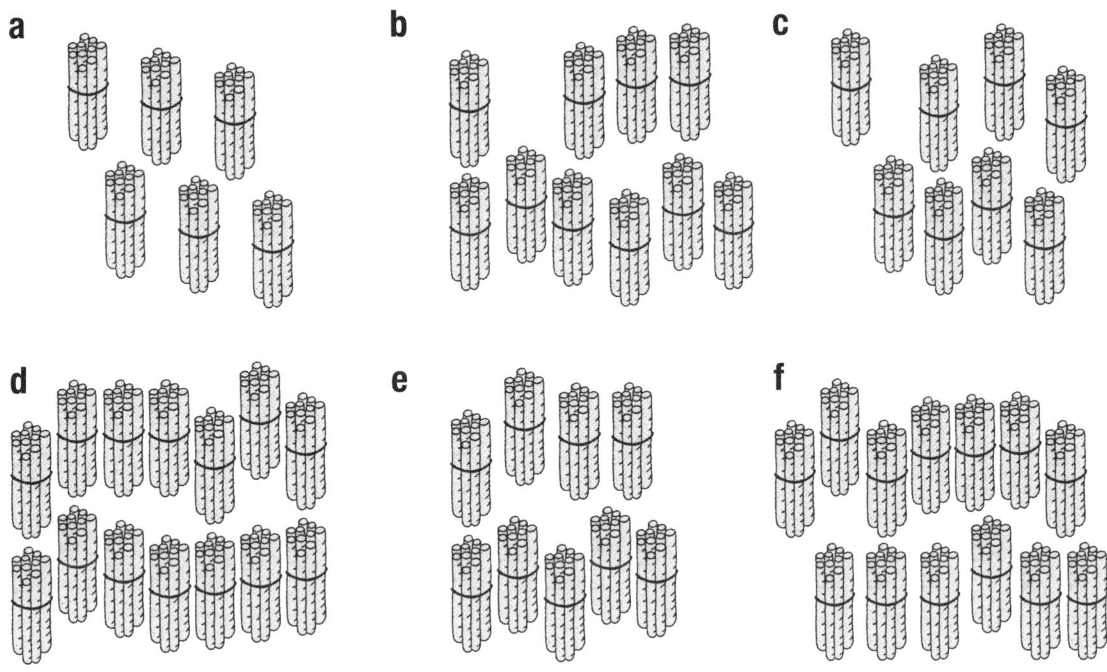

2 Count by 10s to find how much money is in each bag.

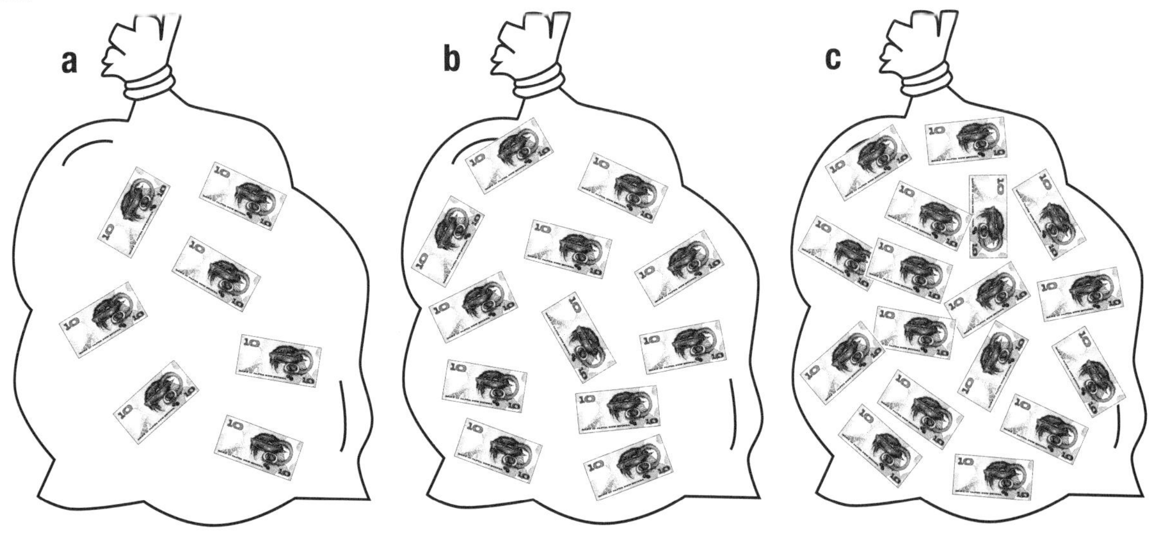

3 Each flower on this page has 10 petals. Count the number of petals on:

a 9 flowers **b** 14 flowers **c** 6 flowers **d** 11 flowers

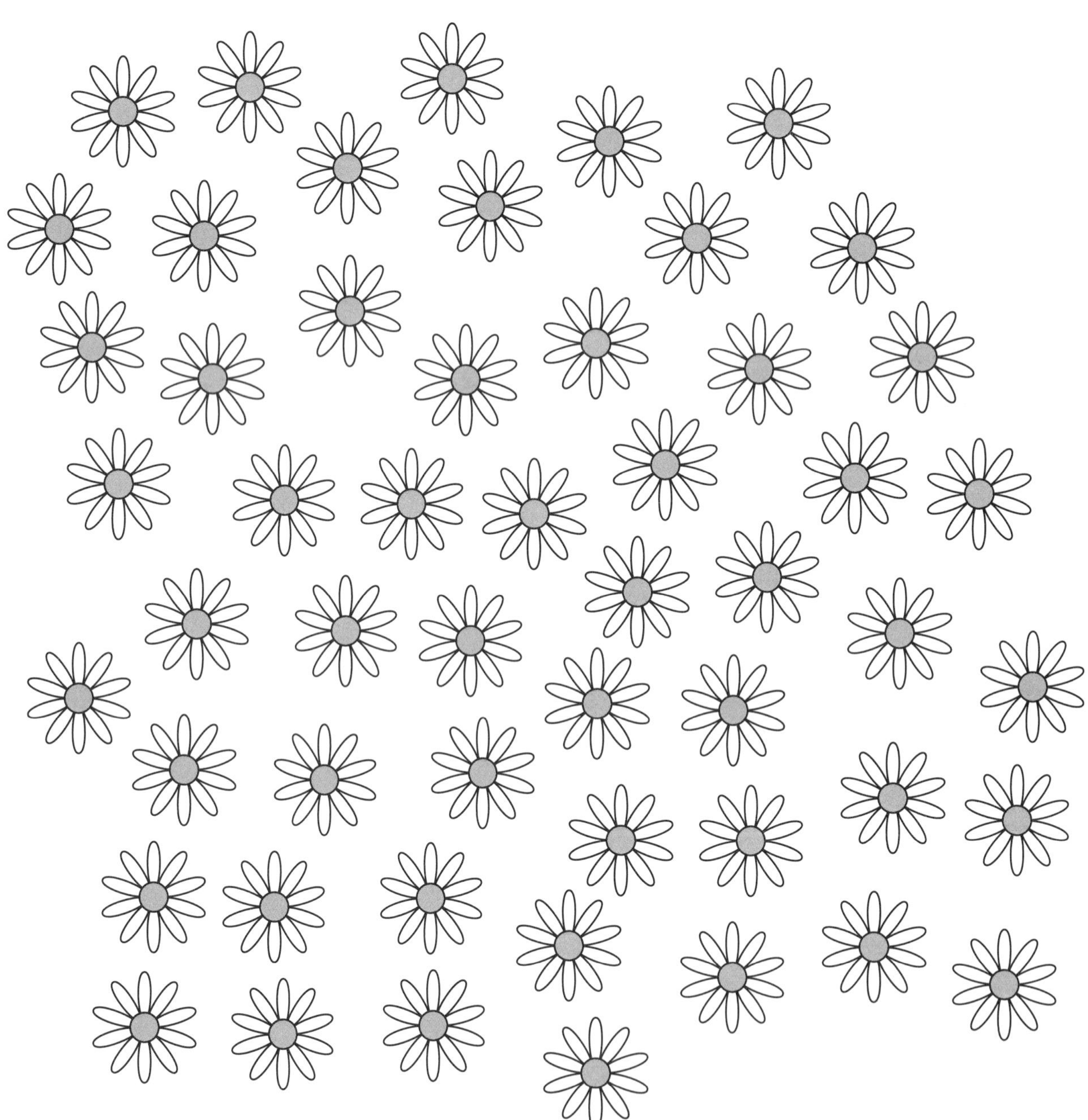

4 Count the number of petals on this page.

Count by 20s

Each group has 20 stars. Count the groups to find the total number of stars on the page.

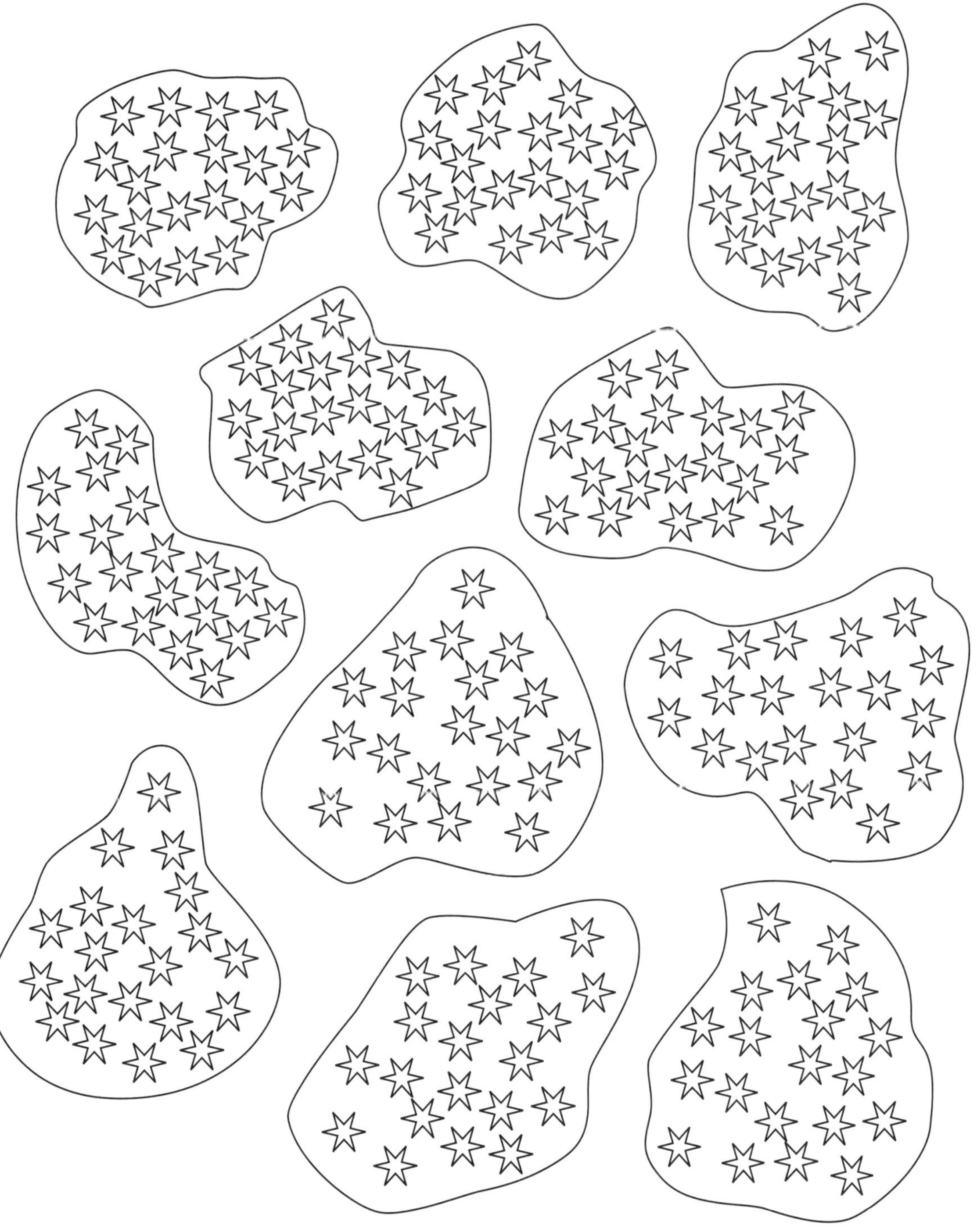

Complete counting sequences by 5s, 10s and 20s

Copy and complete these counting sequences.

1. 44, 49, 54, 59, ____, ____, ____, ____, ____, ____
2. 23, 33, 43, ____, ____, ____, ____, ____, ____
3. 58, 63, 68, ____, ____, ____, ____, ____, ____
4. 91, 81, 71, ____, ____, ____, ____, ____, ____
5. 25, 45, 65, ____, ____, ____, ____, ____, ____
6. 62, 57, 52, ____, ____, ____, ____, ____, ____
7. 220, 200, 180, ____, ____, ____, ____, ____, ____
8. 87, 97, 107, ____, ____, ____, ____, ____, ____
9. 154, 174, 194, ____, ____, ____, ____, ____, ____
10. 336, 326, 316, ____, ____, ____, ____, ____, ____
11. 123, 118, 113, ____, ____, ____, ____, ____, ____
12. 234, 244, 254, ____, ____, ____, ____, ____, ____
13. 549, 529, 509, ____, ____, ____, ____, ____, ____
14. 187, 192, 197, ____, ____, ____, ____, ____, ____
15. 356, 376, 396, ____, ____, ____, ____, ____, ____
16. 448, 438, 428, ____, ____, ____, ____, ____, ____
17. 121, 116, 111, ____, ____, ____, ____, ____, ____
18. 643, 663, 683, ____, ____, ____, ____, ____, ____

Compare groups of 5s

Estimate the larger number of fingers in each pair and then count by 5s to see if you were correct.

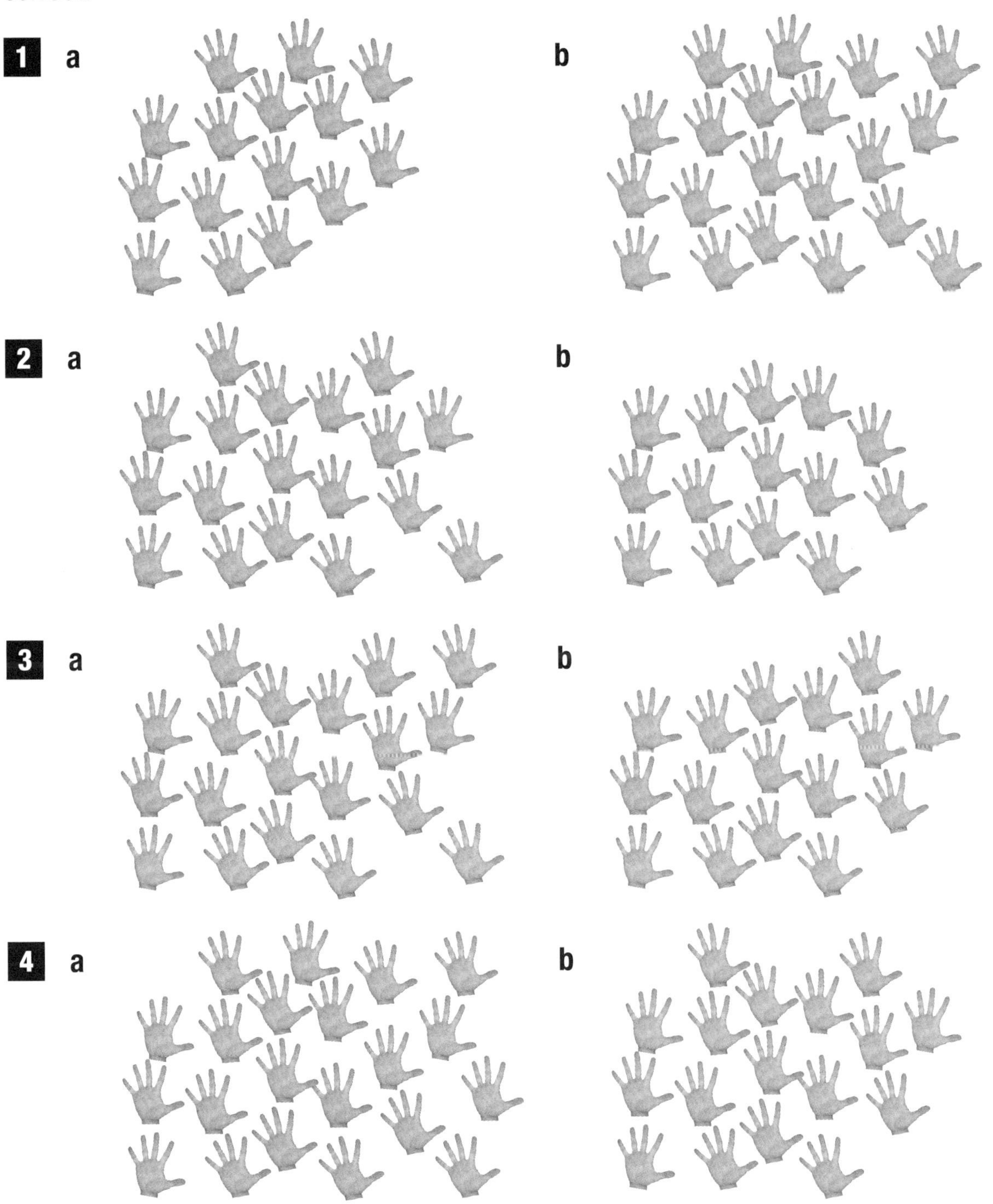

Compare groups of 10s

Estimate the larger number of sticks in each pair and then count by 10s to see if you were correct.

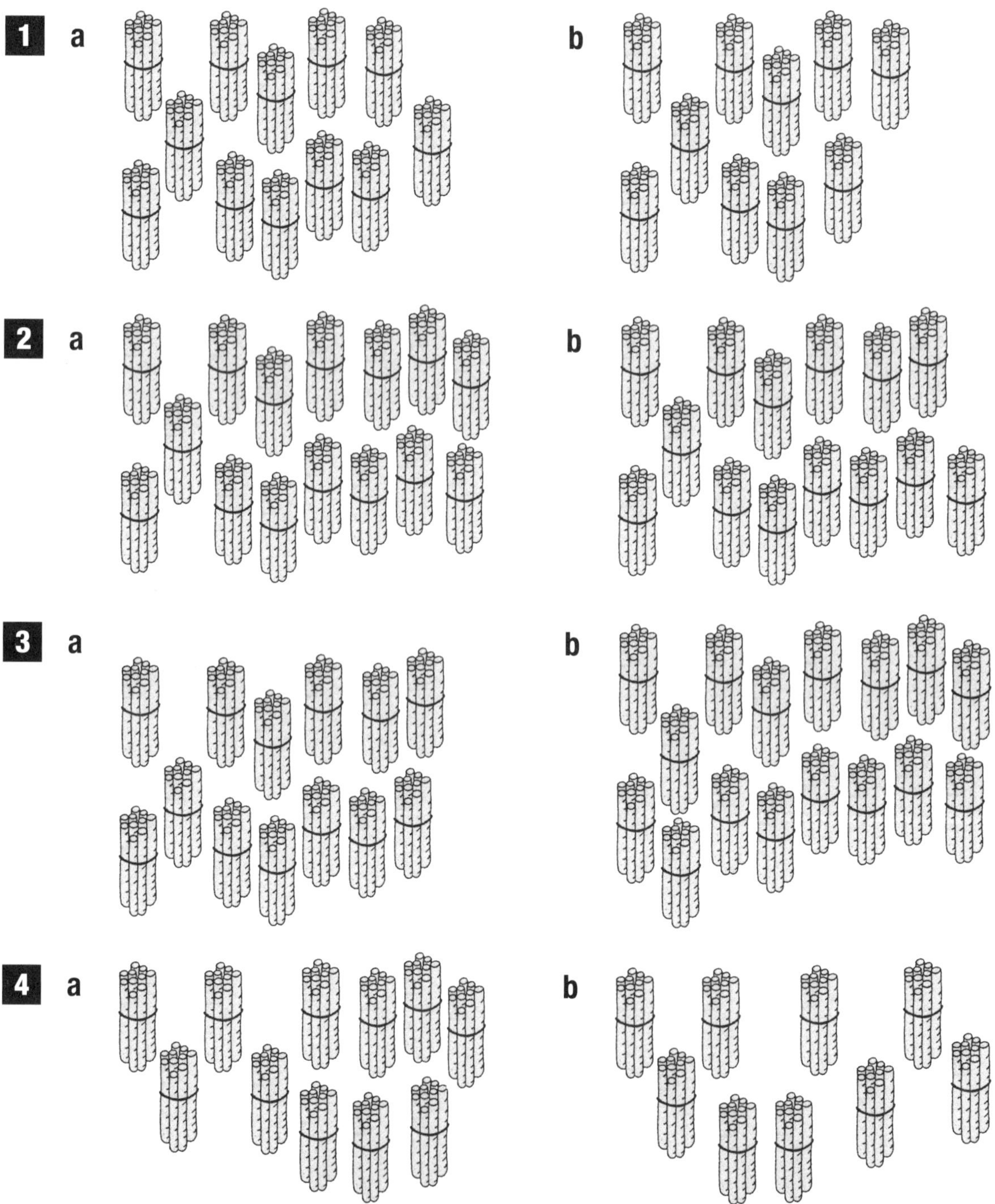

Identify odd and even numbers

1 Write the even numbers from the box below.

63 14 81 35 48
72 90 29 6 34
55 24 17 83 66

2 Write the odd numbers from the box below.

80 25 47 16 54
31 79 50 83 67
95 11 44 58 29

3 Write the even numbers from the box below.

143 420 118 396 105
266 170 274 500 317
209 122 381 238 494

4 Write the odd numbers from the box below.

511 248 383 700 407
113 695 274 829 523
799 466 185 907 616

Predict and calculate odd and even numbers

1 Look at each addition and subtraction then predict if the answer will be odd or even.

a 8 + 9	**b** 6 + 7	**c** 5 + 5	**d** 7 + 9
e 8 – 5	**f** 11 – 6	**g** 14 – 8	**h** 16 – 10
i 5 + 8 + 6	**j** 9 + 5 + 4	**k** 6 + 7 + 5	**l** 12 + 4 + 8
m 13 – 7	**n** 17 – 9	**o** 16 – 7	**p** 14 – 5
q 7 + 7 + 9	**r** 8 + 8 + 7	**s** 9 + 9 + 5	**t** 11 + 5 + 6
u 15 – 8	**v** 13 – 6	**w** 12 – 7	**x** 16 – 8

2 Work out each example to see if you were correct.

3 Predict if the answers to these more difficult additions and subtractions will be odd or even.

a 17 + 15	**b** 27 – 16	**c** 24 + 16	**d** 32 – 18
e 21 – 8	**f** 25 + 37	**g** 24 – 15	**h** 17 + 26
i 15 + 23	**j** 31 – 17	**k** 28 + 26	**l** 43 – 24
m 37 – 19	**n** 25 + 27	**o** 34 – 18	**p** 26 + 33
q 35 + 28	**r** 41 – 16	**s** 35 + 29	**t** 42 – 24

4 Work out each example to see if you were correct.

5 Record five additions and five subtractions of your own that have:

a even answers **b** odd answers

Apply and use the four operations to do calculations with two and three-digit numbers

Rearrange numbers to make them easier to add

1 Write each group of numbers so they are easy to add, and then work out the answers.

a	6	8	5	5	2	4	**b**	7	8	2	4	3	10
c	4	5	9	6	1	4	**d**	5	7	8	5	2	5
e	9	3	8	9	7	1	**f**	8	6	7	4	8	2
g	12	15	5	7	8	13	**h**	4	14	17	6	16	3
i	11	15	2	9	18	7	**j**	10	13	6	9	14	7

2 Add these numbers in your head. Look for ways to make it easier.

a 2 + 5 + 5 + 8 **b** 9 + 10 + 8 + 2 **c** 1 + 1 + 9 + 9

d 4 + 4 + 6 + 6 + 5 **e** 7 + 6 + 5 + 3 + 4 **f** 2 + 5 + 9 + 5 + 1

g 2 + 8 + 1 + 6 + 9 **h** 5 + 3 + 6 + 4 + 5 **i** 3 + 8 + 7 + 5 + 2

j 3 + 2 + 7 + 8 + 9 **k** 5 + 8 + 5 + 5 + 5 **l** 1 + 8 + 2 + 5 + 5

m 8 + 7 + 5 + 2 + 3 **n** 9 + 1 + 4 + 5 + 6

3 Add these larger numbers in your head. Look for ways to make it easier.

a 25 + 70 + 68 + 30 + 75 + 32 **b** 55 + 80 + 16 + 45 + 20 + 4

c 60 + 65 + 22 + 78 + 40 + 35 **d** 15 + 50 + 62 + 38 + 5 + 50

Add and subtract multiples of 10

1 Complete these number sentences.

a $6 + 3 = \square$

$60 + 30 = \square$

b $9 - 4 = \square$

$90 - 40 = \square$

c $4 + 5 = \square$

$40 + 50 = \square$

d $7 - 3 = \square$

$70 - 30 = \square$

2 Fill in the missing numbers.

a $7 - 5 = \square$

$70 - 50 = \square$

b $\square - 3 = 5$

$\square - 30 = 50$

c $2 + \square = 8$

$20 + \square = 80$

d $1 + \square = 8$

$10 + \square = 80$

3 Complete these additions.

a $20 + 50 + 10 = \square$

b $10 + 40 + 30 = \square$

c $30 + 30 + 40 = \square$

d $40 + 10 + 60 = \square$

e $50 + 20 + 80 = \square$

f $30 + 20 + 60 = \square$

g $50 + 30 + 40 = \square$

h $40 + 70 + 30 = \square$

4 Solve these subtractions.

a $80 - 40 = \square$

b $90 - 60 = \square$

c $60 - 50 = \square$

d $70 - 10 = \square$

e $50 - 20 = \square$

f $90 - 30 = \square$

g $60 - 40 = \square$

h $100 - 60 = \square$

Add and subtract 10

Copy each question and fill in the spaces by adding 10 after the number and subtracting 10 before the number. The first one is done for you.

1	57	67	77	87	97	107	117
2	____	____	____	63	____	____	____
3	____	____	____	120	____	____	____
4	____	____	____	151	____	____	____
5	____	____	____	205	____	____	____
6	____	____	____	185	____	____	____
7	____	____	____	92	____	____	____
8	____	____	____	116	____	____	____
9	____	____	____	332	____	____	____
10	____	____	____	284	____	____	____
11	____	____	____	78	____	____	____
12	____	____	____	139	____	____	____
13	____	____	____	460	____	____	____

Use doubles and near doubles when adding mentally

1 Complete these additions.

a $2 + 2 = \square$

$2 + 3 = \square$

$20 + 20 = \square$

$20 + 30 = \square$

b $7 + 7 = \square$

$7 + 8 = \square$

$70 + 70 = \square$

$70 + 80 = \square$

c $5 + 5 = \square$

$5 + 6 = \square$

$50 + 50 = \square$

$50 + 60 = \square$

d $8 + 8 = \square$

$8 + 9 = \square$

$80 + 80 = \square$

$80 + 90 = \square$

e $3 + 3 = \square$

$3 + 4 = \square$

$30 + 30 = \square$

$30 + 40 = \square$

f $10 + 10 = \square$

$10 + 11 = \square$

$100 + 100 = \square$

$100 + 110 = \square$

2 Write answers for these near doubles.

a 11 + 12 = **b** 15 + 16 = **c** 19 + 20 = **d** 15 + 14 =

e 12 + 13 = **f** 20 + 21 = **g** 25 + 26 = **h** 31 + 30 =

i 22 + 23 = **j** 36 + 35 = **k** 51 + 50 = **l** 45 + 44 =

m 18 + 19 = **n** 17 + 18 = **o** 41 + 42 = **p** 34 + 33 =

q 24 + 25 = **r** 80 + 81 = **s** 65 + 66 = **t** 27 + 28 =

3 Try these harder ones.

a 57 + 55 = **b** 26 + 28 = **c** 75 + 77 = **d** 34 + 36 =

Add and subtract 9

Remember

One method of adding 9 is to first add 10, and then subtract 1.
Example: For 25 + 9, say 25 + 10 = 35 and then subtract 1 (35 – 1 = 34), so 25 + 9 = 34.

One method of subtracting 9 is to first subtract 10, and then add 1.
Example: For 31 – 9, say 31 – 10 = 21 and then add 1 (21 + 1 = 22), so 31 – 9 = 22.

1 Add 9 to these numbers.

a	36	**b**	57	**c**	28	**d**	43	**e**	65	**f**	72
g	39	**h**	23	**i**	84	**j**	51	**k**	79	**l**	47
m	155	**n**	128	**o**	232	**p**	319	**q**	266	**r**	198
s	172	**t**	365	**u**	287	**v**	334	**w**	103	**x**	249

2 Subtract 9 from these numbers.

a	52	**b**	71	**c**	38	**d**	93	**e**	86	**f**	60
g	75	**h**	47	**i**	24	**j**	63	**k**	95	**l**	56
m	128	**n**	103	**o**	164	**p**	288	**q**	336	**r**	200
s	157	**t**	215	**u**	321	**v**	122	**w**	274	**x**	340

3 Continue these number sequences by adding or subtracting 9.

a 63, 72, 81, ______, ______, ______, ______, ______, ______, ______, ______

b 177, 168, 159, ______, ______, ______, ______, ______, ______, ______, ______

c 215, 224, 233, ______, ______, ______, ______, ______, ______, ______, ______

Add and subtract 11

Remember

One method of adding 11 is to first add 10, and then add another 1.
Example: For 27 + 11, say 27 + 10 = 37 and then add another 1 (37 + 1 = 38), so 27 + 11 = 38.
One method of subtracting 11 is to first subtract 10, and then subtract another 1.
Example: For 38 – 11, say 38 – 10 = 28 and then subtract another 1 (28 – 1 = 27), so 38 – 11 = 27.

1 Add or subtract 11 as shown.

a	25 + 11 =	**b**	73 – 11 =	**c**	34 + 11 =
d	52 – 11 =	**e**	47 + 11 =	**f**	68 – 11 =
g	16 + 11 =	**h**	89 – 11 =	**i**	41 + 11 =
j	93 – 11 =	**k**	55 + 11 =	**l**	70 – 11 =
m	107 + 11 =	**n**	154 – 11 =	**o**	231 + 11 =
p	283 – 11 =	**q**	129 + 11 =	**r**	200 – 11 =
s	178 + 11 =	**t**	342 – 11 =	**u**	216 + 11 =
v	307 – 11 =	**w**	295 + 11 =	**x**	209 – 11 =

2 Continue these number sequences by adding or subtracting 11.

a 58, 69, 80, _____, _____, _____, _____, _____, _____, _____, _____

b 124, 113, 102, _____, _____, _____, _____, _____, _____, _____, _____

c 276, 287, 298, _____, _____, _____, _____, _____, _____, _____, _____

d 213, 202, 191, _____, _____, _____, _____, _____, _____, _____, _____

Add and subtract 9, 10 and 11

1 Copy and complete each grid.

a

+	9	10	11
32			
67			
188			
145			
209			

b

−	9	10	11
84			
46			
117			
231			
183			

2 Copy and complete the number chains.

a 53 + 11 = ☐ − 9 = ☐ + 10 = ☐ − 11 = ☐ + 9 = ☐

b 17 − 9 = ☐ + 10 = ☐ + 11 = ☐ − 9 = ☐ + 11 = ☐

c 89 + 10 = ☐ − 9 = ☐ − 11 = ☐ + 9 = ☐ − 11 = ☐

d 136 + 11 = ☐ + 10 = ☐ − 9 = ☐ − 9 = ☐ + 10 = ☐

e 214 − 10 = ☐ − 9 = ☐ − 11 = ☐ + 10 = ☐ − 9 = ☐

f 195 − 11 = ☐ + 9 = ☐ + 11 = ☐ − 9 = ☐ + 10 = ☐

g 241 + 9 = ☐ + 11 = ☐ − 10 = ☐ − 11 = ☐ + 9 = ☐

h 308 + 10 = ☐ − 11 = ☐ + 9 = ☐ − 10 = ☐ − 11 = ☐

Add and subtract 99

Remember

A short method of adding 99 is to first add 100, and then subtract 1.
Example: For 53 + 99, say 53 + 100 = 153 and then subtract 1 (153 – 1 = 152), so 53 + 99 = 152.

A short method of subtracting 99 is to first subtract 100, and then add 1.
Example: For 178 – 99, say 178 – 100 = 78 and then add 1 (78 + 1 = 79), so 178 – 99 = 79.

1 Add 99 as shown.

- **a** 76 + 99 =
- **b** 59 + 99 =
- **c** 65 + 99 =
- **d** 82 + 99 =
- **e** 34 + 99 =
- **f** 28 + 99 =
- **g** 47 + 99 =
- **h** 188 + 99 =
- **i** 225 + 99 =
- **j** 156 + 99 =
- **k** 475 + 99 =
- **l** 384 + 99 =
- **m** 527 + 99 =
- **n** 314 + 99 =

2 Subtract 99 as shown.

- **a** 164 – 99 =
- **b** 281 – 99 =
- **c** 388 – 99 =
- **d** 217 – 99 =
- **e** 128 – 99 =
- **f** 462 – 99 =
- **g** 353 – 99 =
- **h** 529 – 99 =
- **i** 436 – 99 =
- **j** 335 – 99 =
- **k** 151 – 99 =
- **l** 730 – 99 =
- **m** 504 – 99 =
- **n** 631 – 99 =

Add and subtract single digits

Complete each number chain to show how to get from one number to the next. The first one has been done for you.

1 15, 17, 16, 20, 15

(+2) (−1) (+4) (−5)

2 12, 14, 10, 8, 11

3 10, 6, 5, 10, 20

4 18, 15, 17, 20, 25

5 20, 14, 12, 17, 15

6 14, 17, 19, 12, 16

7 19, 25, 20, 18, 22

8

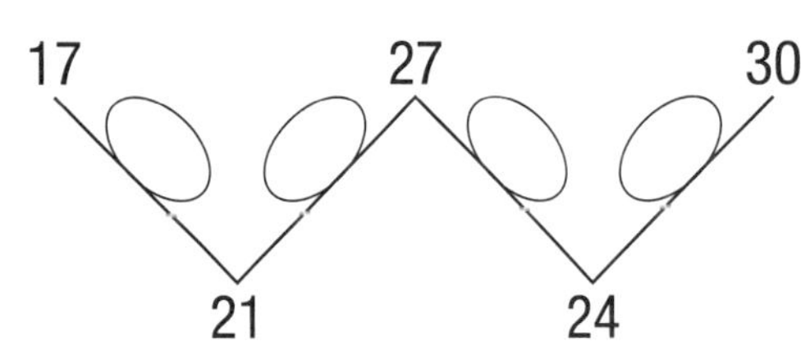

9 11, 8, 15, 19, 13

10

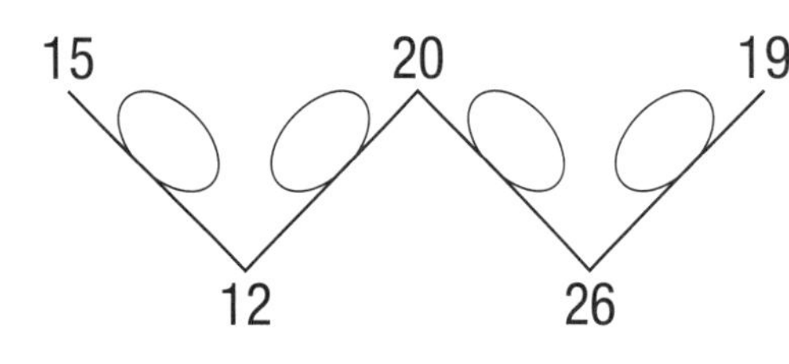

11

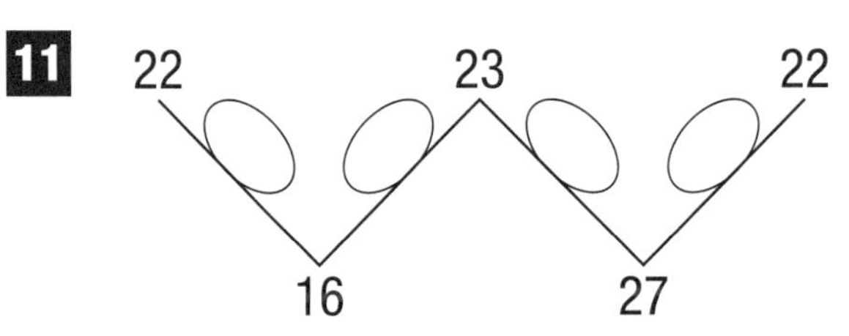

12

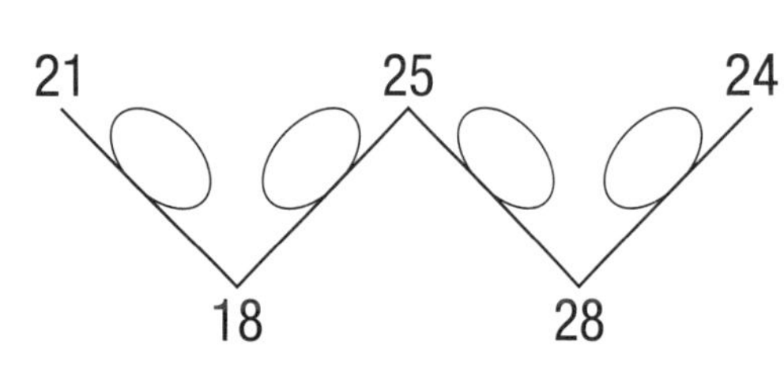

Make number pairs for 100 and 1000

1 Copy and complete this chart.

30	90	60	80	70	50	40	10
+	+	+	+	+	+	+	+
100	100	100	100	100	100	100	100

2 If you have the number shown, how much more is needed to make 100?

a 60	**b** 30	**c** 90	**d** 40	**e** 50	**f** 20
g 65	**h** 58	**i** 73	**j** 51	**k** 46	**l** 87
m 39	**n** 78	**o** 24	**p** 32	**q** 43	**r** 25
s 18	**t** 69	**u** 27	**v** 71	**w** 56	**x** 38

3 Match a number on the top line with a number on the bottom line to make 1000.

200	700	950	840	510	750
490	250	300	50	800	160

4 If you have the number shown, how much more is needed to make 1000?

a 400	**b** 150	**c** 350	**d** 280	**e** 190	**f** 460
g 880	**h** 570	**i** 610	**j** 930	**k** 420	**l** 760
m 805	**n** 595	**o** 755	**p** 285	**q** 915	**r** 375

Find missing numbers to 20

1 Copy and complete each addition.

a	6 + □ = 15	**b**	□ + 7 = 14	**c**	□ + 3 = 12
d	8 + □ = 15	**e**	□ + 6 = 14	**f**	3 + □ = 13
g	7 + □ = 16	**h**	□ + 4 = 11	**i**	□ + 9 = 15
j	9 + □ = 13	**k**	□ + 8 = 11	**l**	□ + 10 = 19
m	6 + □ = 12	**n**	4 + □ = 12	**o**	5 + □ = 16
p	□ + 7 = 15	**q**	7 + □ = 13	**r**	8 + □ = 17
s	□ + 8 = 14	**t**	□ + 9 = 18	**u**	8 + □ = 16
v	□ + 9 = 20	**w**	5 + □ = 11	**x**	7 + □ = 17

2 Copy and complete each subtraction.

a	16 – □ = 7	**b**	□ – 6 = 14	**c**	□ – 4 = 12
d	15 – □ = 9	**e**	17 – □ = 12	**f**	12 – □ = 7
g	□ – 8 = 4	**h**	□ – 5 = 9	**i**	14 – □ = 6
j	□ – 7 = 13	**k**	□ – 3 = 15	**l**	11 – □ = 5
m	13 – □ = 7	**n**	□ – 6 = 11	**o**	14 – □ = 8
p	□ – 9 = 9	**q**	□ – 7 = 10	**r**	18 – □ = 12
s	15 – □ = 8	**t**	□ – 8 = 3	**u**	19 – □ = 13
v	□ – 4 = 14	**w**	□ – 5 = 11	**x**	13 – □ = 4

Find the difference

1 Write the pairs of numbers that have a difference of 5.

a 76 | 81

b 65 | 71

c 38 | 42

d 98 | 103

e 54 | 57

f 60 | 65

g 45 | 49

h 58 | 63

i 79 | 84

j 30 | 36

k 22 | 27

l 120 | 125

m 19 | 28

n 167 | 169

o 92 | 97

p 236 | 241

q 159 | 164

r 185 | 192

s 108 | 113

t 200 | 205

u 145 | 150

v 212 | 218

w 183 | 188

x 132 | 142

y 121 | 125

z 109 | 112

2 Write the pairs of numbers that have a difference of 50.

a | 10 | 60

b | 50 | 100

c

d | 70 | 110

e | 20 | 70

f | 60 | 110

g | 100 | 150

h | 30 | 60

i | 5 | 55

j | 20 | 80

k | 200 | 250

l | 80 | 130

m | 70 | 120

n | 100 | 200

o | 300 | 350

p | 15 | 60

q | 190 | 240

r | 25 | 75

s | 45 | 95

t | 550 | 600

u | 75 | 100

v | 95 | 145

w | 210 | 260

x | 35 | 80

y | 390 | 440

z | 120 | 180

Use mental strategies for addition

Play this game with a friend. You need two dice and each player needs 15 counters (each player's set a different colour).

Take turns to toss the dice together and find the total. Cover that number on the game board with your counter. If the total of your throw is already covered, you miss that turn. When all numbers on the board are covered, the winner is the player with the most counters on the board.

5	12	4	8
7	3	6	10
2	9	7	8
10	5	11	6

Use mental strategies for subtraction

Play this game with one or two friends. You need two sets of cards numbered 1 to 12. Each player needs a game board and nine counters.

Shuffle the cards and place them face down in a pile. Take turns to pick up the top two cards from the pile and find the difference between the two numbers. If the difference is on the game board, cover it with a counter. Continue until one player has covered all numbers on their game board. This player is the winner.

6	8	1
0	3	5
4	10	2

Practise the two times table

1 Complete the 2× tables.

a $4 \times 2 =$ ______ **b** $8 \times 2 =$ ______ **c** $6 \times 2 =$ ______

d $5 \times 2 =$ ______ **e** $11 \times 2 =$ ______ **f** $9 \times 2 =$ ______

g $3 \times 2 =$ ______ **h** $7 \times 2 =$ ______ **i** $10 \times 2 =$ ______

j $1 \times 2 =$ ______ **k** $12 \times 2 =$ ______ **l** $2 \times 2 =$ ______

2 Use the answers from the previous activity to help you complete these divisions.

a $20 \div 2 =$ ______ **b** $6 \div 2 =$ ______ **c** $10 \div 2 =$ ______

d $16 \div 2 =$ ______ **e** $8 \div 2 =$ ______ **f** $14 \div 2 =$ ______

g $2 \div 2 =$ ______ **h** $24 \div 2 =$ ______ **i** $4 \div 2 =$ ______

j $12 \div 2 =$ ______ **k** $22 \div 2 =$ ______ **l** $18 \div 2 =$ ______

3 Fill the gaps.

a $\square \times 2 = 18$ **b** $\square \times 2 = 6$ **c** $\square \times 2 = 12$

d $\square \times 2 = 4$ **e** $\square \times 2 = 16$ **f** $\square \times 2 = 10$

g $\square \times 2 = 8$ **h** $\square \times 2 = 20$ **i** $\square \times 2 = 14$

4 Double each number.

a 5 ______ **b** 8 ______ **c** 6 ______ **d** 7 ______

e 11 ______ **f** 9 ______ **g** 12 ______ **h** 4 ______

Practise the five times table

1 Complete the multiplication wheels.

a

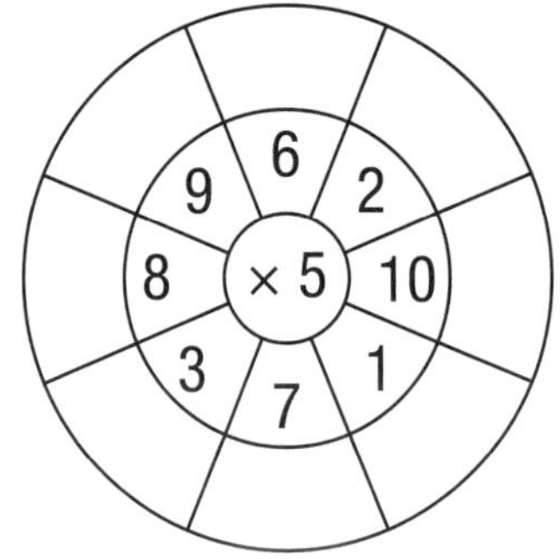

b

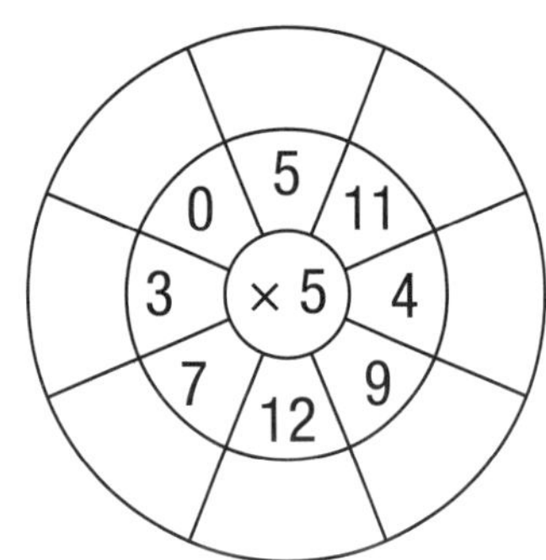

2 Fill the gaps.

a $\square \times 5 = 25$ **b** $10 \times 5 = \square$ **c** $5 \times \square = 15$

d $\square \times 5 = 40$ **e** $\square \times 5 = 30$ **f** $\square \times 5 = 5$

g $11 \times 5 = \square$ **h** $45 = \square \times 5$ **i** $4 \times 5 = \square$

j $\square \times 5 = 35$ **k** $5 \times \square = 60$ **l** $10 = 5 \times \square$

3 Write a 5× table for each answer.

a $\square \times \square = 20$ **b** $\square \times \square = 45$ **c** $\square \times \square = 10$

d $\square \times \square = 30$ **e** $\square \times \square = 15$ **f** $\square \times \square = 25$

g $\square \times \square = 40$ **h** $\square \times \square = 60$ **i** $\square \times \square = 35$

4 Use the answers from the previous activity to help you complete these divisions.

a $25 \div 5 =$ ______ **b** $40 \div 5 =$ ______ **c** $15 \div 5 =$ ______

d $50 \div 5 =$ ______ **e** $35 \div 5 =$ ______ **f** $20 \div 5 =$ ______

Practise the ten times table

1 Copy and complete the multiplication grid.

×	5	8	2	6	0	9	1	3	7	10	4
10											

2 Use the answers from the previous activity to help you complete these divisions.

a $90 \div 10 =$ ______ **b** $40 \div 10 =$ ______ **c** $10 \div 10 =$ ______

d $70 \div 10 =$ ______ **e** $20 \div 10 =$ ______ **f** $50 \div 10 =$ ______

g $100 \div 10 =$ ______ **h** $60 \div 10 =$ ______ **i** $30 \div 10 =$ ______

3 Write a 10× table for each answer.

a ☐ × ☐ = 100 **b** ☐ × ☐ = 60 **c** ☐ × ☐ = 80

d ☐ × ☐ = 0 **e** ☐ × ☐ = 30 **f** ☐ × ☐ = 120

g ☐ × ☐ = 50 **h** ☐ × ☐ = 70 **i** ☐ × ☐ = 20

j ☐ × ☐ = 90 **k** ☐ × ☐ = 40 **l** ☐ × ☐ = 110

4 Complete the multiplication and division wheels.

a

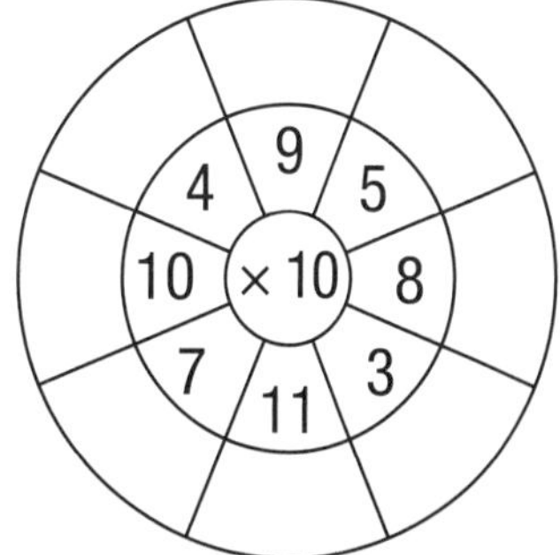

b

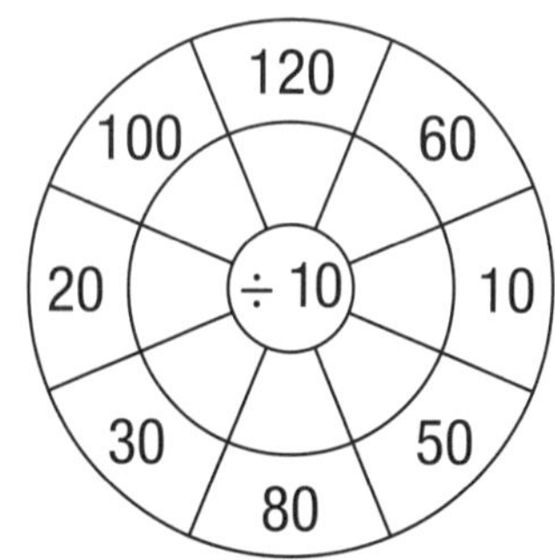

Practise the one, two, five and ten times tables

1 Copy and complete the multiplication grid.

×	4	1	9	7	0	6	2	8	3	10	5
1											
2											
5											
10											

2 Fill the gaps.

a $2 \times 9 = \square$ **b** $\square \times 10 = 60$ **c** $5 \times \square = 30$

d $10 \times \square = 80$ **e** $2 \times 5 = \square$ **f** $\square \times 5 = 40$

g $5 \times 5 = \square$ **h** $\square \times 6 = 30$ **i** $4 \times 10 = \square$

j $\square \times 5 = 35$ **k** $7 \times \square = 7$ **l** $9 \times 5 = \square$

m $8 \times \square = 16$ **n** $\square \times 4 = 20$ **o** $\square \times 10 = 70$

p $7 \times 2 = \square$ **q** $4 \times 2 = \square$ **r** $5 \times \square = 55$

s $\square \times 12 = 12$ **t** $3 \times \square = 15$ **u** $10 \times 11 = \square$

3 Fill the gaps with a 2×, 5× or 10× table so the answer is correct.

a $\square \times \square = 30$ **b** $\square \times \square = 15$ **c** $\square \times \square = 24$

d $\square \times \square = 20$ **e** $\square \times \square = 45$ **f** $\square \times \square = 18$

g $\square \times \square = 25$ **h** $\square \times \square = 60$ **i** $\square \times \square = 14$

Add two-digit numbers

Help Box

Add whole numbers without trading:

52 + 43

$$\begin{array}{r} 52 \\ +\ 43 \\ \hline 95 \end{array}$$

Add whole numbers with trading:

46 + 58

$$\begin{array}{r} 46 \\ +\ {}_{1}58 \\ \hline 104 \end{array}$$

1 How many eggs can you collect?

a $\begin{array}{r} 63 \\ +\ 24 \\ \hline \end{array}$

b $\begin{array}{r} 45 \\ +\ 51 \\ \hline \end{array}$

c $\begin{array}{r} 27 \\ +\ 62 \\ \hline \end{array}$

d $\begin{array}{r} 33 \\ +\ 46 \\ \hline \end{array}$

e $\begin{array}{r} 47 \\ +\ 45 \\ \hline \end{array}$

f $\begin{array}{r} 57 \\ +\ 36 \\ \hline \end{array}$

g $\begin{array}{r} 69 \\ +\ 35 \\ \hline \end{array}$

h $\begin{array}{r} 64 \\ +\ 49 \\ \hline \end{array}$

i $\begin{array}{r} 75 \\ +\ 57 \\ \hline \end{array}$

j $\begin{array}{r} 84 \\ +\ 65 \\ \hline \end{array}$

k $\begin{array}{r} 35 \\ +\ 76 \\ \hline \end{array}$

l $\begin{array}{r} 96 \\ +\ 85 \\ \hline \end{array}$

How many eggs did you collect?

2 Add the amounts of money to find the bag that has the most.

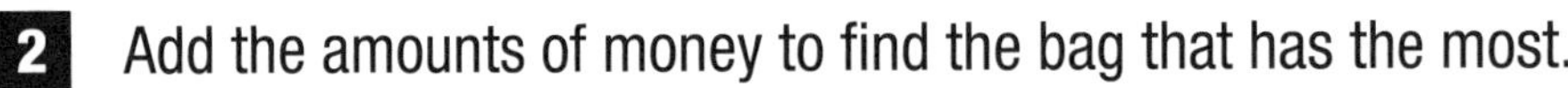

a

b

c

d

e

f

g

h

i

j

k

l

Which bag has the most?

Add three-digit numbers

Help Box

Add whole numbers without trading:

243 + 355

$$\begin{array}{r} 2\,4\,3 \\ +\,3\,5\,5 \\ \hline 5\,9\,8 \\ \hline \end{array}$$

Add whole numbers with trading:

378 + 163

$$\begin{array}{r} 3\,7\,8 \\ +\,{}_{1}1\,{}_{1}6\,3 \\ \hline 5\,4\,1 \\ \hline \end{array}$$

How many bubbles can you burst?

1

$$\begin{array}{r} 325 \\ +163 \\ \hline \end{array}$$

2

$$\begin{array}{r} 253 \\ +226 \\ \hline \end{array}$$

3

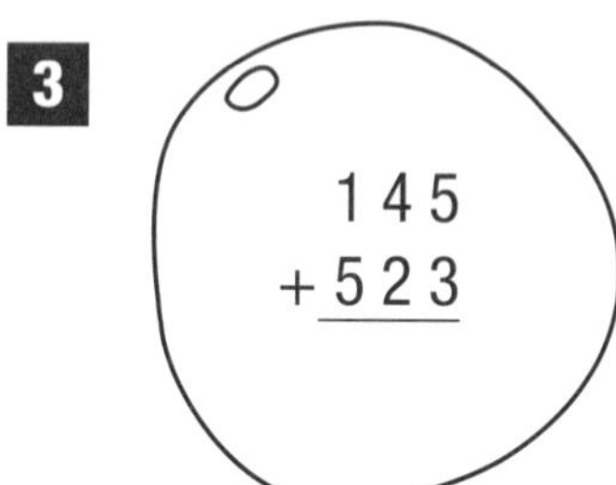

$$\begin{array}{r} 145 \\ +523 \\ \hline \end{array}$$

4

$$\begin{array}{r} 339 \\ +452 \\ \hline \end{array}$$

5

$$\begin{array}{r} 267 \\ +573 \\ \hline \end{array}$$

6

$$\begin{array}{r} 686 \\ +345 \\ \hline \end{array}$$

7

$$\begin{array}{r} 458 \\ +366 \\ \hline \end{array}$$

8

$$\begin{array}{r} 396 \\ +267 \\ \hline \end{array}$$

9

$$\begin{array}{r} 557 \\ +165 \\ \hline \end{array}$$

How many bubbles did you burst?

Add two and three-digit numbers

Find the answers to crack the code.

1

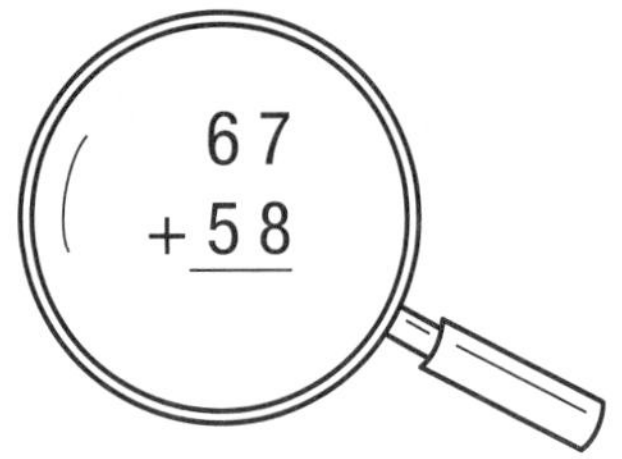

2

3

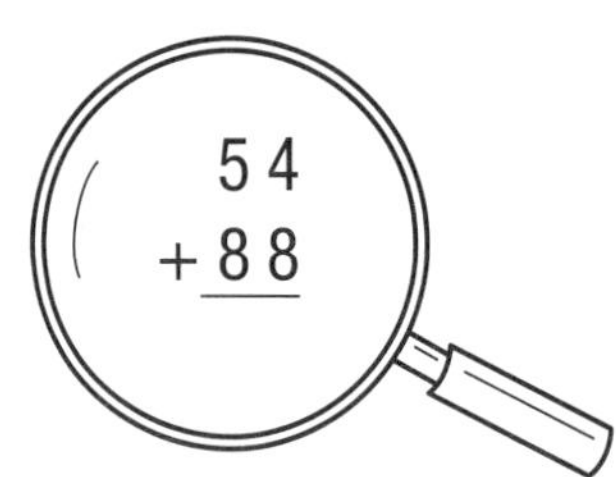

4

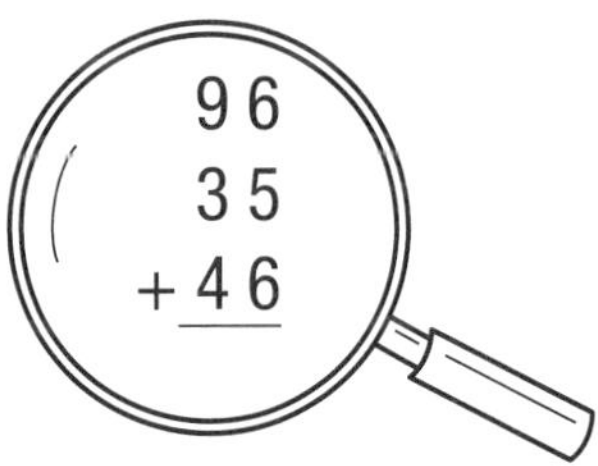

5

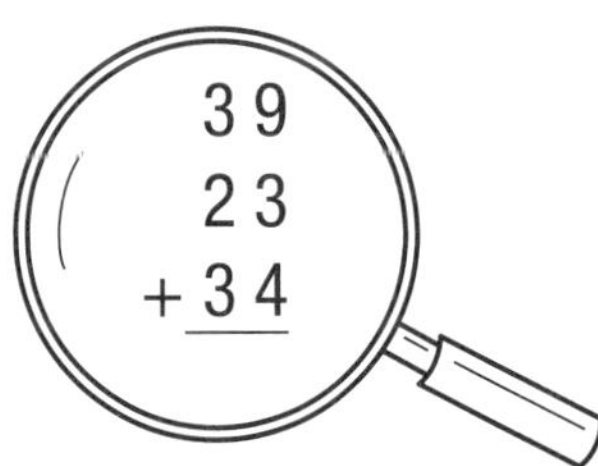

6

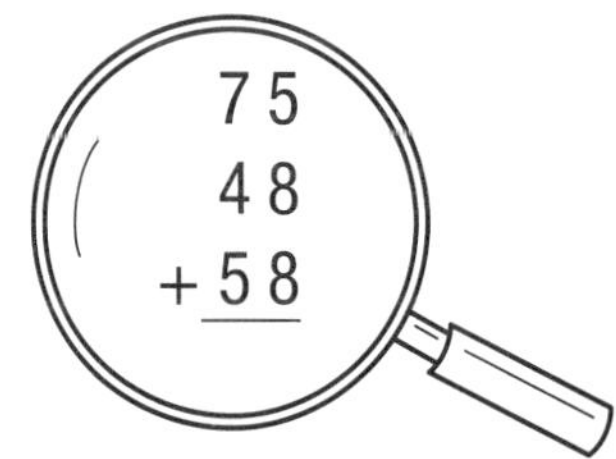

7

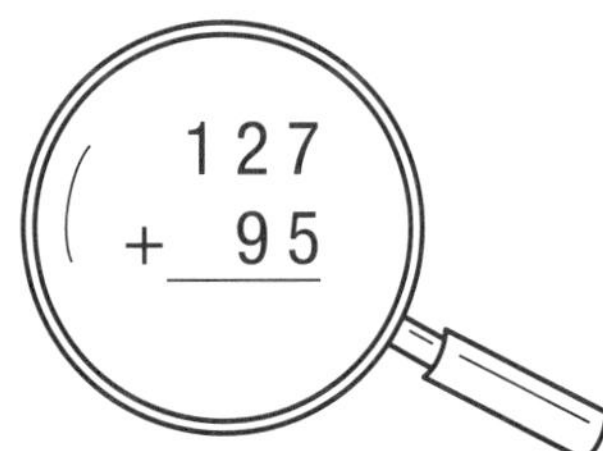

8

9

10

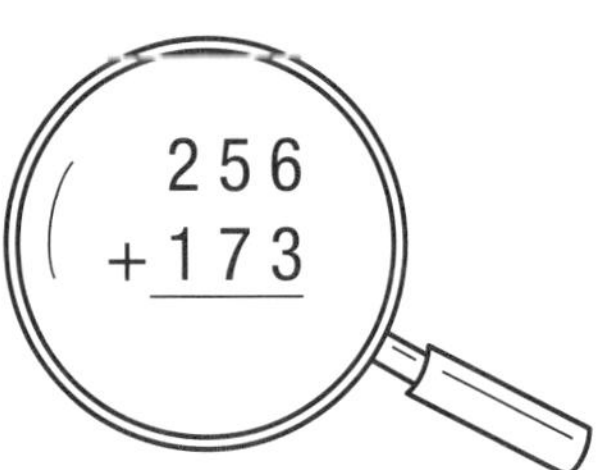

11

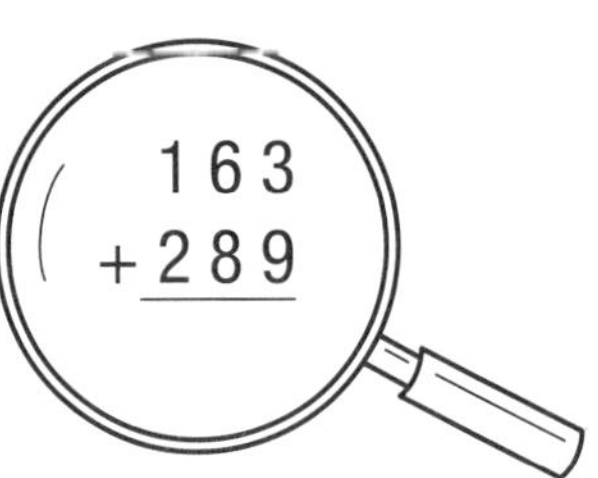

12

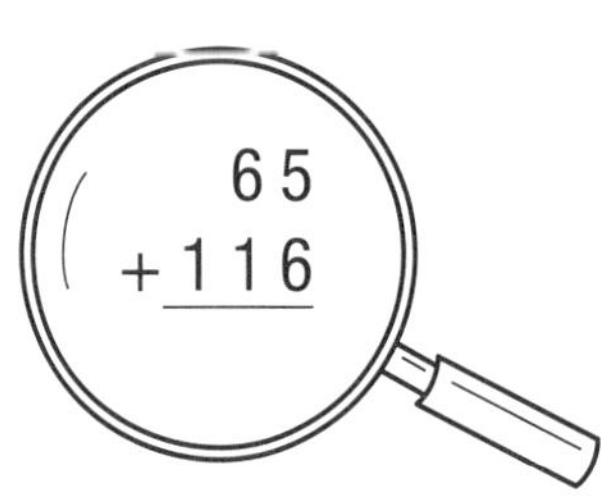

L = 222 E = 181 V = 125 R = 142 Y = 177 W = 96 D = 451 O = 429 N = 452

____ ____ ____ ____ ____ ____ ____ ____ ____ ____ ____ ____

Solve addition word problems

Set out and solve each addition.

1 A trade store owner sold 76 bags of rice in March and 58 bags of rice in April. How many bags of rice did he sell during the two months?

2 One village has 117 people and another village has 85 people. What is the total number of people in both villages?

3 Violet has K138 in her bank account. If she puts in another K83, how much will be in her account?

4 At the market, 127 kilograms of potatoes were sold one week and 96 kilograms the next week. How many kilograms of potatoes were sold for the two weeks?

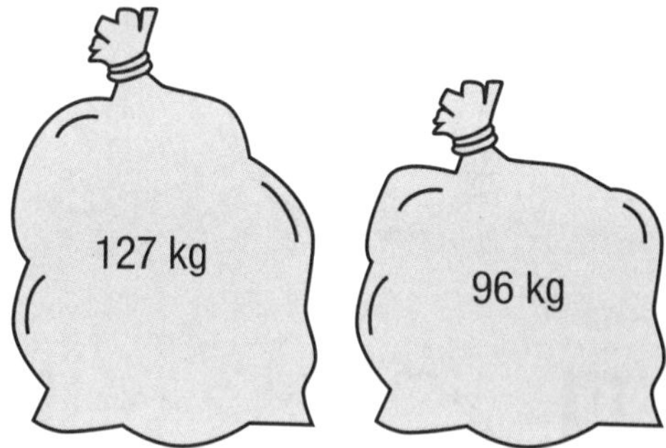

5 A local two-day show had 267 visitors on the first day and 358 visitors on the second day. How many people visited the show over the two days?

6 A group of villagers planted 83 trees on Monday, 105 trees on Tuesday and 76 trees on Wednesday. How many trees were planted over the three days?

7 A concert had 247 spectators on its opening night and 276 on its closing night. What was the total number of spectators for the two nights?

Subtract two-digit numbers

Help Box

Subtract whole numbers without trading:

78 – 45

$$\begin{array}{r} 7\,8 \\ -\ 4\,5 \\ \hline 3\,3 \end{array}$$

Subtract whole numbers with trading:

81 – 35

$$\begin{array}{r} {}^{7}\not{8}\,{}^{1}1 \\ -\ 3\,5 \\ \hline 4\,6 \end{array}$$

1 Solve each subtraction to find out which cars are breaking the speed limit.

a

b

c

d

e

f

g

h

i

j

k

2 How many balloons can you burst?

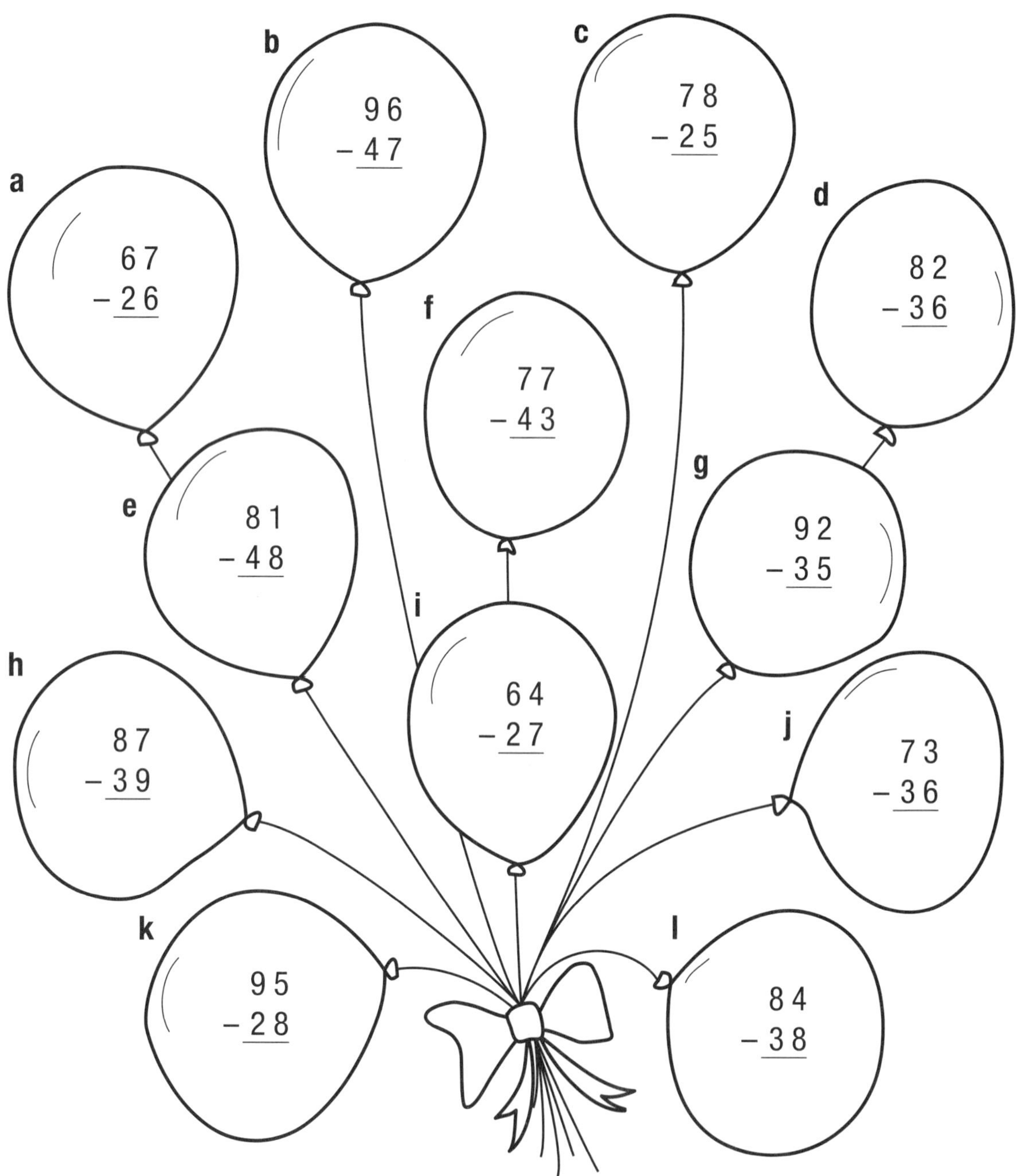

How many balloons did you burst?

Subtract three-digit numbers

Help Box

Subtract whole numbers without trading:

265 – 141

$$\begin{array}{r} 265 \\ -\,141 \\ \hline 124 \\ \hline \end{array}$$

Subtract whole numbers with trading:

342 – 175

$$\begin{array}{r} {}^{2}\not{3}\,{}^{13}\not{4}\,{}^{1}2 \\ -\,1\;\,7\;\,5 \\ \hline 1\;\,6\;\,7 \\ \hline \end{array}$$

How many fish can you catch?

1

2

3

4

5

6

7

8

9

How many fish did you catch?

Subtract two and three-digit numbers

Solve the subtractions to keep the boats afloat.

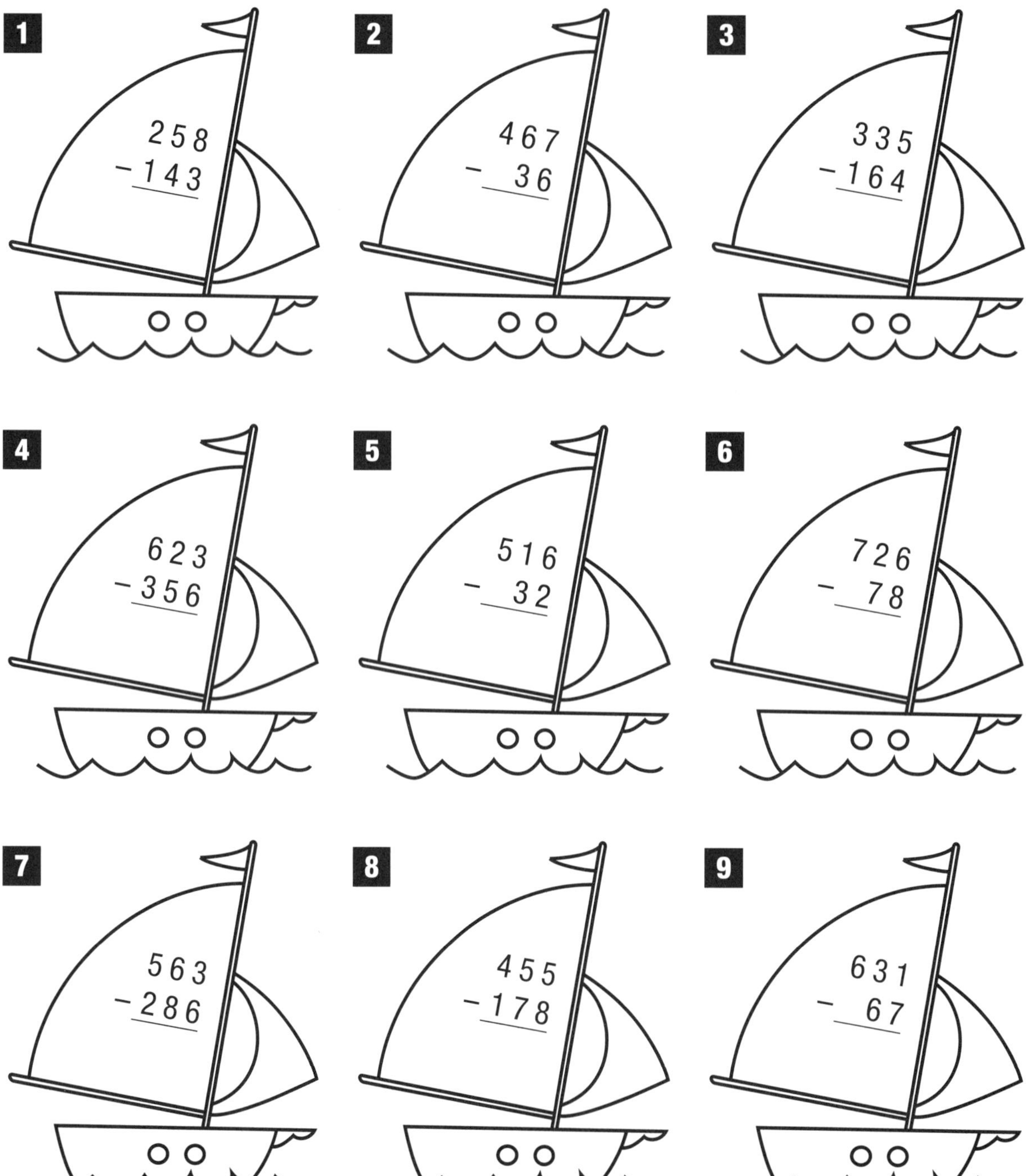

Solve subtraction word problems

Set out and solve each subtraction.

1. Bexley has K433 in his bank account. If he withdraws K78 to pay for car repairs, how much will still be in his account?

2. A hotel has 217 people booked for Saturday night and 138 booked for Sunday night. How many more people are booked for Saturday night than Sunday night?

3. Some villagers had 165 trees to plant. If they have already planted 86 trees, how many are left to plant?

4. The total distance we have to travel is 352 kilometres. If we have travelled 174 kilometres, how much further do we have to travel?

5. Lillian needs K416 for a plane ticket. She has saved K267. How much more does she need?

6. Of the 283 people to visit the health clinic, 158 have been immunised. How many people still have to be immunised?

7. A truck was carrying goods with a total weight of 521 kilograms. After delivering goods weighing 266 kilograms, what weight of goods was still in the truck?

Understand multiplication as equal groupings and repeated addition

Write an addition and a multiplication number sentence for each picture. For example, the first one is 13 + 13 + 13 = 39 and 3 × 13 = 39.

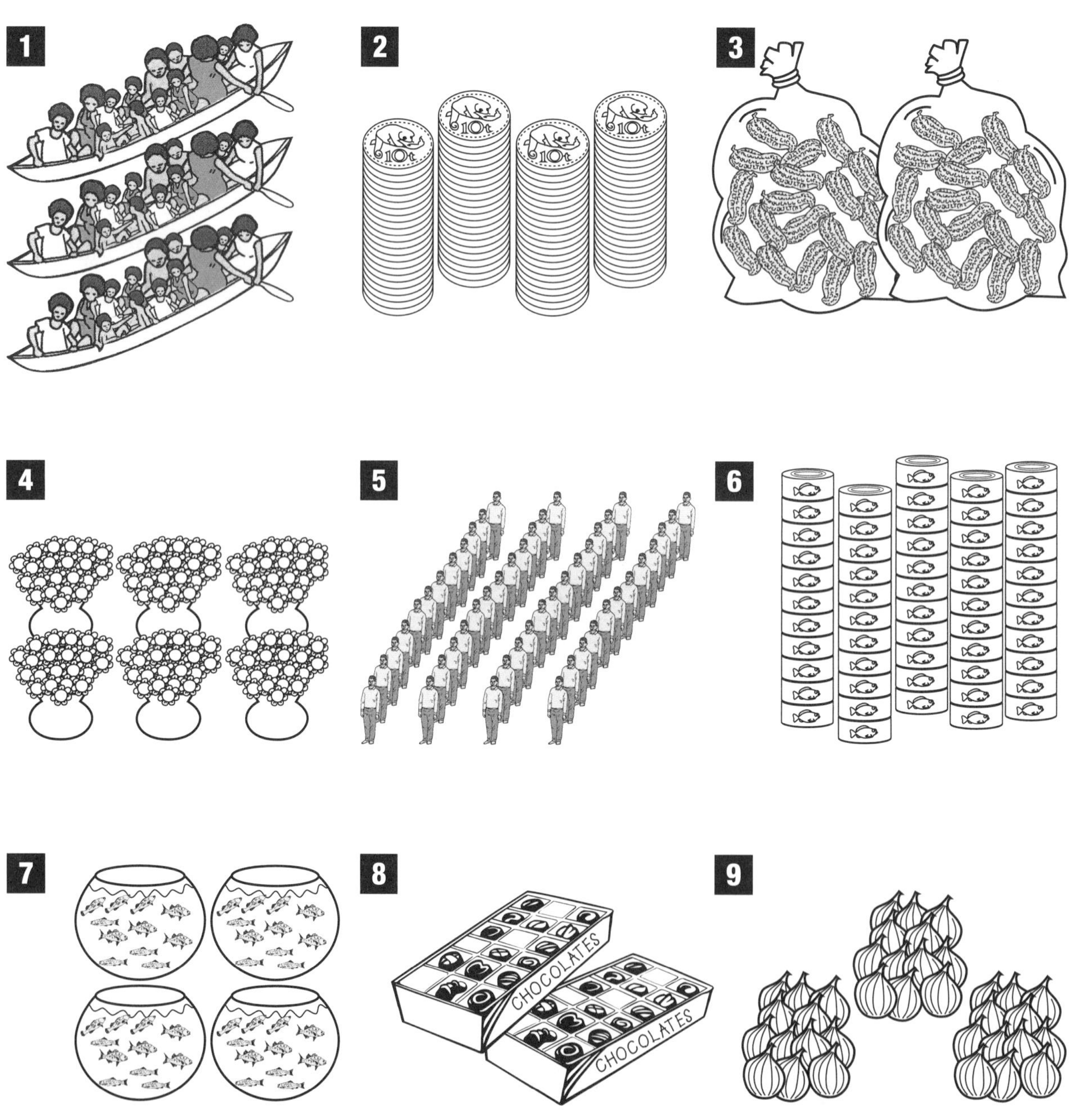

Multiply two-digit numbers

Help Box

Multiply whole numbers without trading:

$$\begin{array}{r} 32 \\ \times\ 4 \\ \hline 128 \end{array}$$

Multiply whole numbers with trading:

$$\begin{array}{r} {}^{2}5\,6 \\ \times\ 4 \\ \hline 224 \end{array}$$

1 Each correct answer saves a frog from the crocodile.
How many frogs will your crocodile have for dinner?

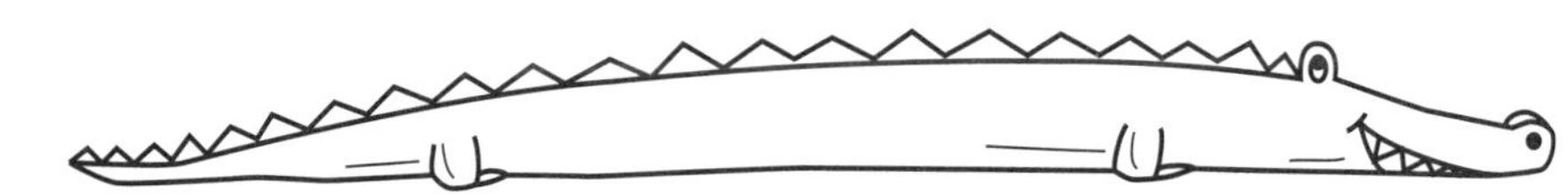

a

b

c

d

e

f

g

h

2 Set out and solve these multiplications.

a $\begin{array}{r} 26 \\ \times\ 3 \\ \hline \end{array}$ **b** $\begin{array}{r} 18 \\ \times\ 5 \\ \hline \end{array}$ **c** $\begin{array}{r} 16 \\ \times\ 6 \\ \hline \end{array}$ **d** $\begin{array}{r} 43 \\ \times\ 4 \\ \hline \end{array}$ **e** $\begin{array}{r} 75 \\ \times\ 3 \\ \hline \end{array}$ **f** $\begin{array}{r} 64 \\ \times\ 4 \\ \hline \end{array}$

g $\begin{array}{r} 55 \\ \times\ 3 \\ \hline \end{array}$ **h** $\begin{array}{r} 37 \\ \times\ 5 \\ \hline \end{array}$ **i** $\begin{array}{r} 28 \\ \times\ 6 \\ \hline \end{array}$ **j** $\begin{array}{r} 45 \\ \times\ 7 \\ \hline \end{array}$ **k** $\begin{array}{r} 34 \\ \times\ 4 \\ \hline \end{array}$ **l** $\begin{array}{r} 53 \\ \times\ 6 \\ \hline \end{array}$

Match and solve multiplication word problems

Write the multiplication number sentence that matches each word problem, and then solve it.

1 Yerema bought 6 concert tickets at K27 each. What was the total cost of the tickets?

$$\begin{array}{r} 18 \\ \times\ 7 \\ \hline \end{array}$$

2 Four trucks each carried 48 boxes. How many boxes altogether?

$$\begin{array}{r} 36 \\ \times\ 4 \\ \hline \end{array}$$

3 There are 24 drinks in each carton. How many drinks are there in 5 cartons?

$$\begin{array}{r} 24 \\ \times\ 5 \\ \hline \end{array}$$

4 Each suitcase weighs 36 kilograms. What is the total weight of 4 suitcases?

$$\begin{array}{r} 27 \\ \times\ 6 \\ \hline \end{array}$$

5 There are 56 potatoes in each bag. How many potatoes are in 3 bags?

$$\begin{array}{r} 45 \\ \times\ 5 \\ \hline \end{array}$$

6 Each packet contains 18 biscuits. How many biscuits are in 7 packets?

$$\begin{array}{r} 48 \\ \times\ 4 \\ \hline \end{array}$$

7 Daniel earns K32 each week at a part-time job. How much will he have earned after 8 weeks?

$$\begin{array}{r} 56 \\ \times\ 3 \\ \hline \end{array}$$

8 Florence travels 45 kilometres each day. How far does she travel in 5 days?

$$\begin{array}{r} 32 \\ \times\ 8 \\ \hline \end{array}$$

Relate division to multiplication

1 Match each multiplication fact with its related division fact.

a $8 \times 2 = 16$ **b** $5 \times 7 = 35$ **c** $6 \times 5 = 30$ **d** $3 \times 8 = 24$

e $4 \times 9 = 36$ **f** $8 \times 10 = 80$ **g** $4 \times 4 = 16$ **h** $6 \times 3 = 18$

$36 \div 9 = 4$	$30 \div 6 = 5$	$35 \div 5 = 7$	$16 \div 4 = 4$
$80 \div 10 = 8$	$16 \div 8 = 2$	$18 \div 6 = 3$	$24 \div 3 = 8$

2 Use the multiplication facts to help you work out the answer to each division fact.

a $4 \times 8 = 32$, so $32 \div 4 = \square$ and $32 \div 8 = \square$

b $7 \times 9 = 63$, so $63 \div 9 = \square$ and $63 \div 7 = \square$

c $3 \times 4 = 12$, so $12 \div 4 = \square$ and $12 \div 3 = \square$

d $5 \times 8 = 40$, so $40 \div 5 = \square$ and $40 \div 8 = \square$

e $7 \times 3 = 21$, so $21 \div 3 = \square$ and $21 \div 7 = \square$

f $9 \times 3 = 27$, so $27 \div 3 = \square$ and $27 \div 9 = \square$

g $2 \times 11 = 22$, so $22 \div 11 = \square$ and $22 \div 2 = \square$

h $5 \times 3 = 15$, so $15 \div 3 = \square$ and $15 \div 5 = \square$

i $7 \times 2 = 14$, so $14 \div 7 = \square$ and $14 \div 2 = \square$

j $4 \times 5 = 20$, so $20 \div 4 = \square$ and $20 \div 5 = \square$

Understand division as equal groupings and repeated subtraction

1 Use the pictures to help you make equal groups and then write a division number sentence for each picture. For example, the first one is 16 ÷ 4 = 4.

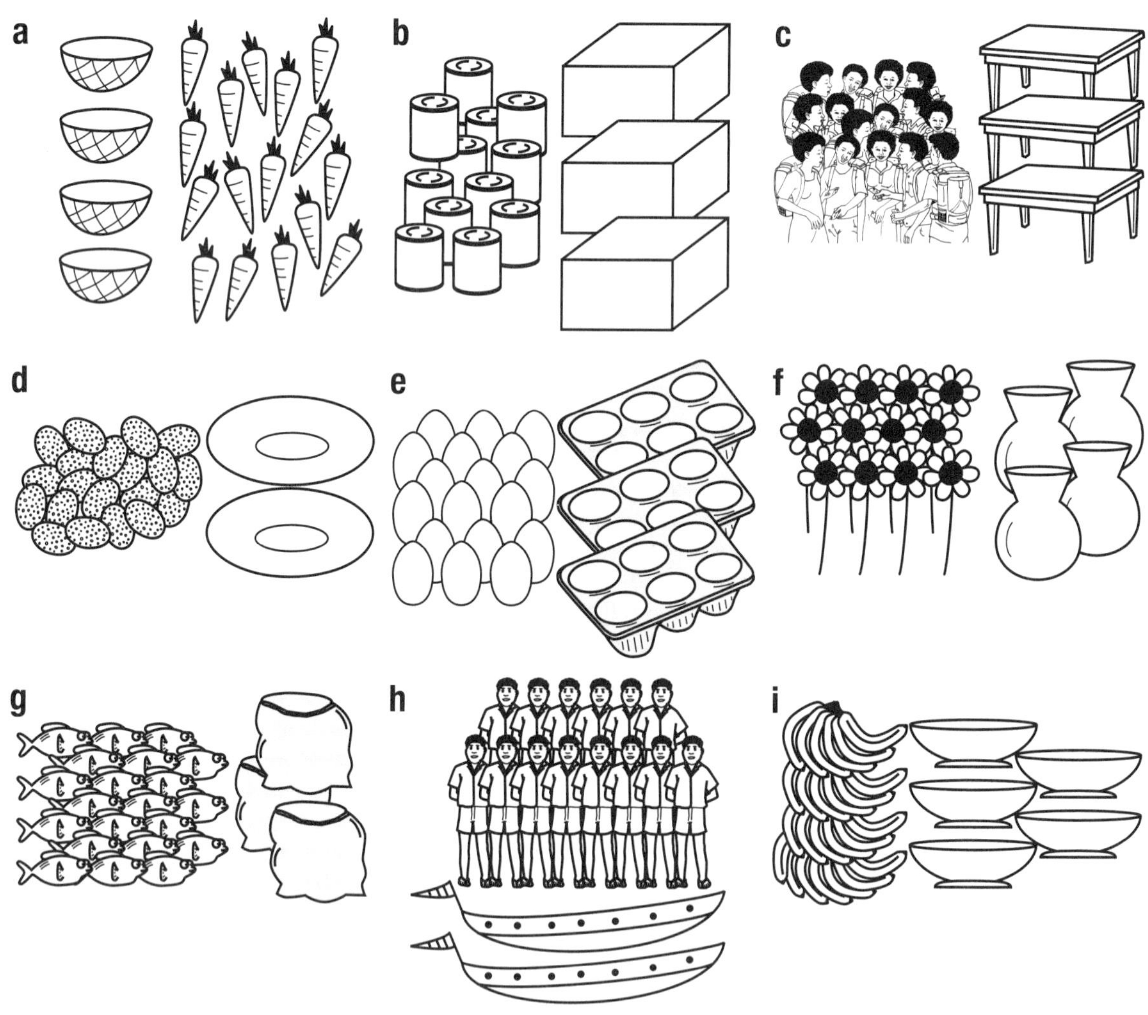

2 Use repeated subtraction to find the answers to these divisions. For example, 24 ÷ 6 = 24 – 6 – 6 – 6 – 6 = 0, so 24 ÷ 6 = 4.

a	48 ÷ 8	**b**	35 ÷ 5	**c**	50 ÷ 10	**d**	36 ÷ 6	**e**	18 ÷ 3	**f**	16 ÷ 4
g	20 ÷ 5	**h**	28 ÷ 7	**i**	30 ÷ 5	**j**	40 ÷ 10	**k**	15 ÷ 3	**l**	25 ÷ 5

Divide with and without remainders

Remember

When the answer to a division number sentence has a remainder, for example 25 ÷ 6, this is how we write it: 25 ÷ 6 = 4 rem. 1

1 Copy and solve these division number sentences.

a	22 ÷ 10 =	**b**	16 ÷ 8 =	**c**	15 ÷ 6 =	**d**	63 ÷ 10 =
e	21 ÷ 2 =	**f**	30 ÷ 5 =	**g**	20 ÷ 3 =	**h**	33 ÷ 5 =
i	27 ÷ 3 =	**j**	42 ÷ 5 =	**k**	32 ÷ 8 =	**l**	29 ÷ 4 =
m	14 ÷ 3 =	**n**	17 ÷ 2 =	**o**	36 ÷ 6 =	**p**	25 ÷ 4 =
q	23 ÷ 5 =	**r**	44 ÷ 10 =	**s**	19 ÷ 6 =	**t**	70 ÷ 10 =
u	55 ÷ 5 =	**v**	31 ÷ 6 =	**w**	37 ÷ 4 =	**x**	35 ÷ 7 =

2 Complete the gaps in these division number sentences.

a	24 ÷ ☐ = 12	**b**	50 ÷ ☐ = 5	**c**	12 ÷ ☐ = 2
d	☐ ÷ 10 = 9	**e**	☐ ÷ 6 = 3	**f**	☐ ÷ 8 = 5
g	27 ÷ ☐ = 5 rem. 2	**h**	63 ÷ ☐ = 10 rem. 3	**i**	38 ÷ ☐ = 6 rem. 2
j	43 ÷ ☐ = 10 rem. 3	**k**	19 ÷ ☐ = 4 rem. 3	**l**	31 ÷ ☐ = 5 rem. 1
m	52 ÷ ☐ = 5 rem. 2	**n**	26 ÷ ☐ = 6 rem. 2	**o**	44 ÷ ☐ = 8 rem. 4
p	39 ÷ ☐ = 6 rem. 3	**q**	15 ÷ ☐ = 2 rem. 1	**r**	34 ÷ ☐ = 4 rem. 6

Practise division with remainders

Play this game with friends (no more than four players). You need two sets of cards numbered 2, 3, 4, 5 and 10, and a different coloured counter for each player.

Shuffle the cards and place them face down. Put your counters on START and when it is your turn, move your counter to number 10.

Take turns to pick up a card and divide the number your counter is on by the number on the card, and move the remainder. For example, *if you are on 11 and pick up the number 3, your answer is 3 remainder 2, so you move forward 2 spaces.*

The first player to reach 50 wins the game.

Start here									10
11	12	13	14	15	16	17	18	19	20
21	22	23	24	25	26	27	28	29	30
31	32	33	34	35	36	37	38	39	40
41	42	43	44	45	46	47	48	49	50

Use multiplication and division to solve word problems

Decide whether to use multiplication or division and then solve each word problem.

1. The dressmaker has 18 buttons. She puts 3 on each dress. How many dresses will she finish?

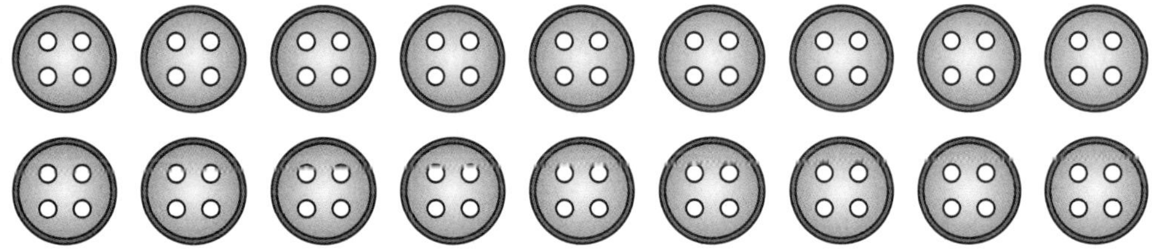

2. There are 4 boxes with 46 books each. How many books are there altogether?

3. Each basket contains 23 pieces of fruit. If there are 6 baskets, how many pieces of fruit are there altogether?

4. A school has 48 tennis balls. If they are shared equally among 6 groups of students, how many balls will each group get?

5. There are 80 books that have to be packed so that there are 10 in each package. How many packages will there be?

6. Each box weighs 47 kilograms. What is the total weight of 5 boxes?

7. The total number of people in 3 boats is 33. If each boat has the same number of people, how many are in each boat?

Use the four operations to solve word problems

Decide what process to use and then solve each word problem.

1. The baker made 125 bread rolls this morning. Now there are 37 left. How many have been sold?

2. Lydia travels the same distance each day and after 5 days finds she has travelled 55 kilometres. How far does she travel each day?

3. On the opening night, 187 people went to the concert. On the second night, 136 people went. How many people attended over the two nights?

4. Joe bought two items at a total cost of K244. One of the items cost K175. How much was the other one?

5. Each container holds 58 litres of water. How many litres of water will 8 containers hold?

6. Nina picks 40 bananas. If she puts 5 in each bag, how many bags will she need?

7. Three suitcases are being loaded into a car. One weighs 27 kilograms, another weighs 36 kilograms and the third weighs 32 kilograms. What is their total weight?

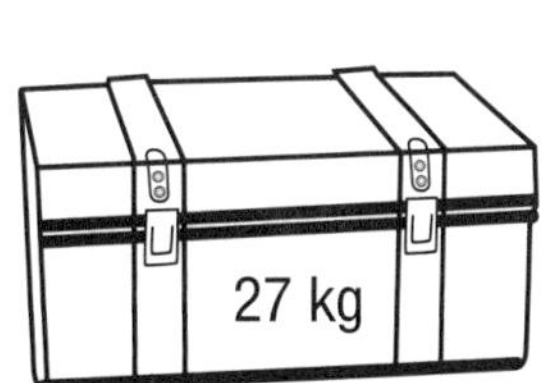

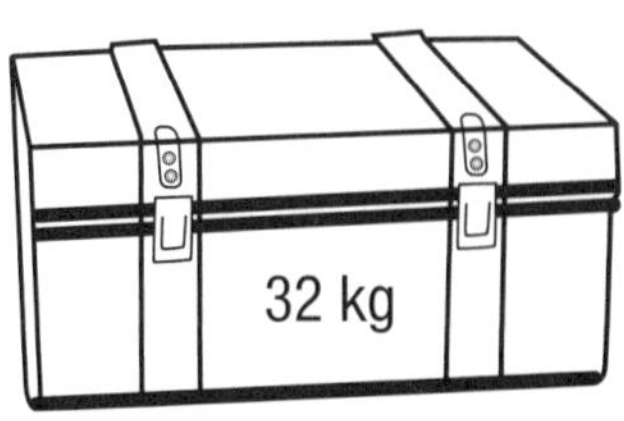

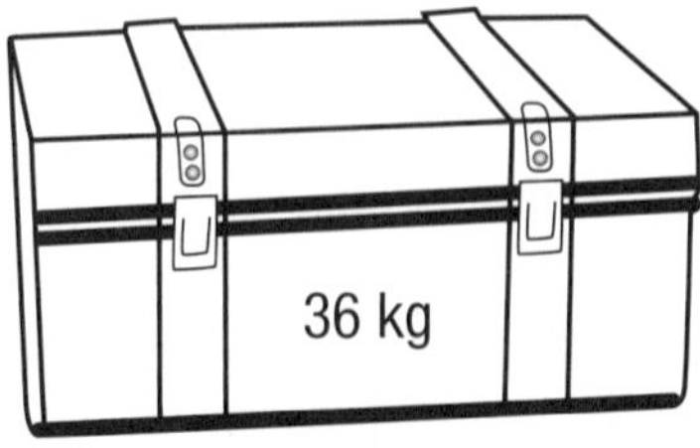

Identify and recognise common fractions

Recognise halves of shapes and groups

Which of these shapes or groups show one half ($\frac{1}{2}$)?

1
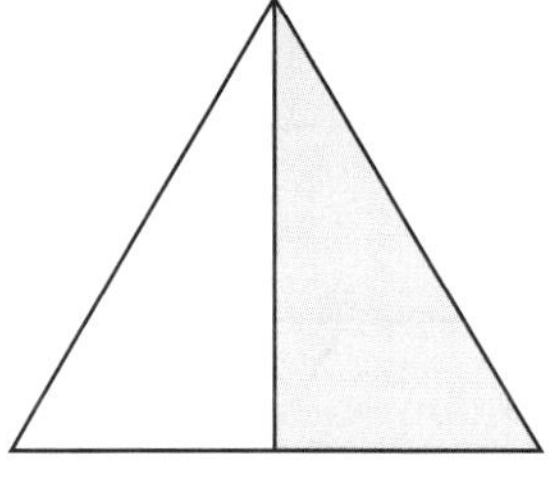

2
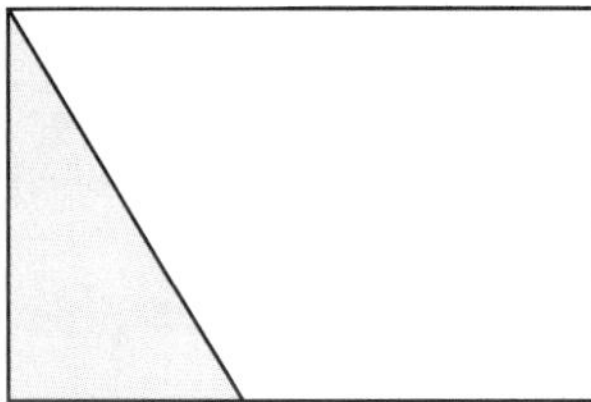

3
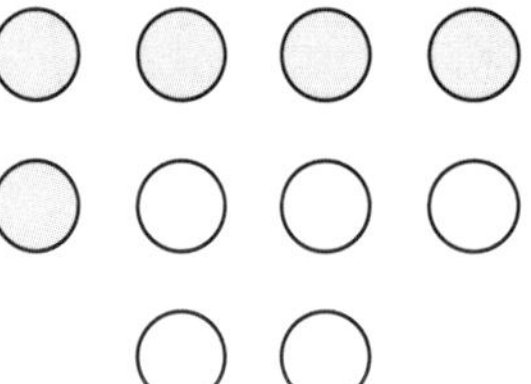

4
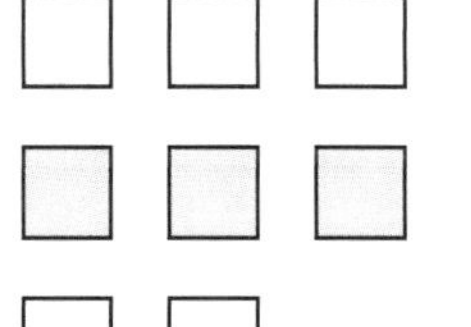

5
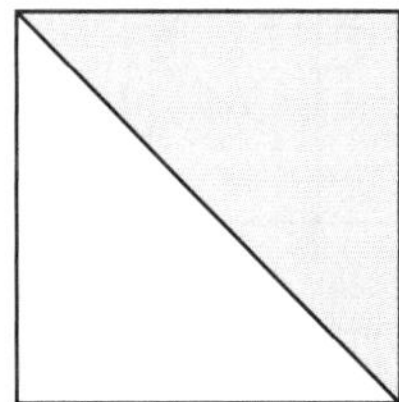

6
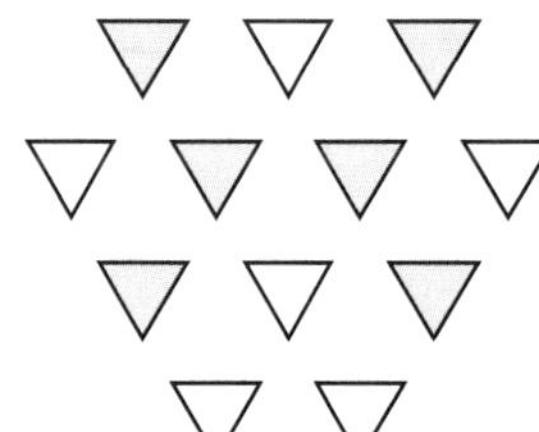

7
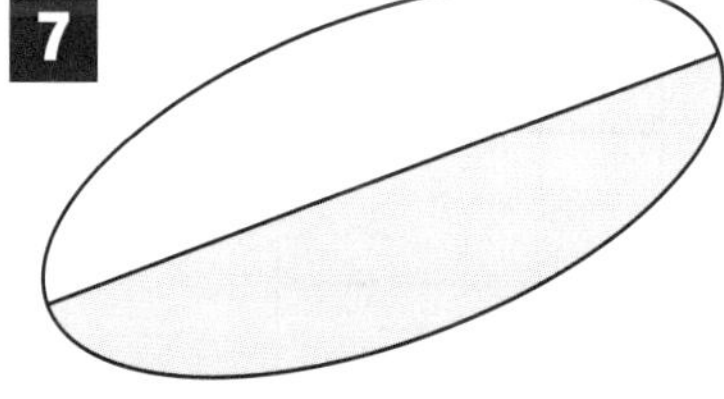

8
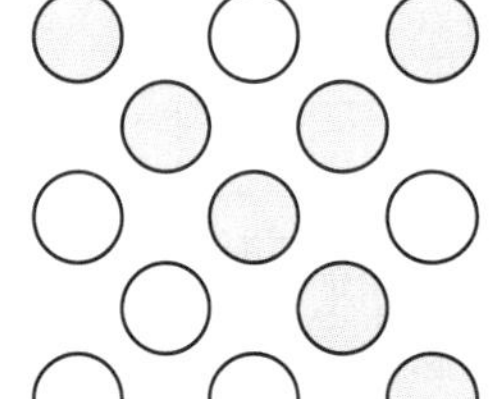

9
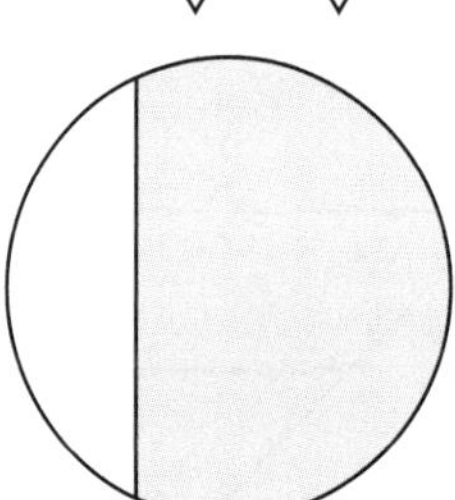

10

11
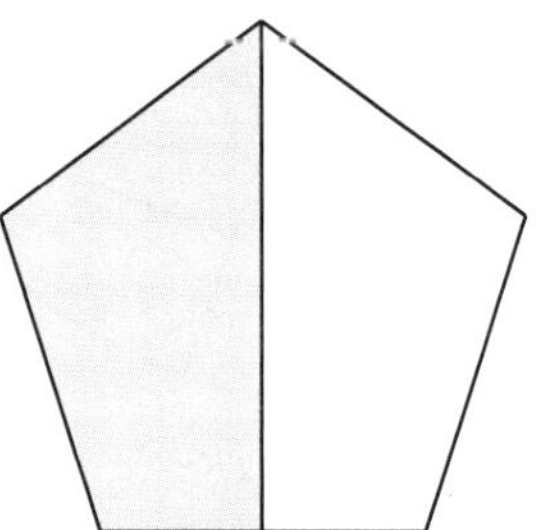

12
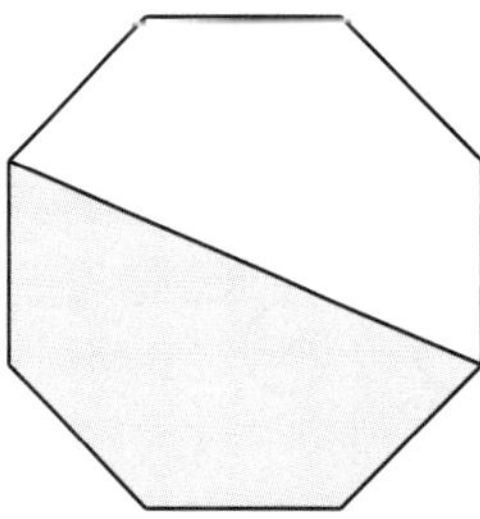

13
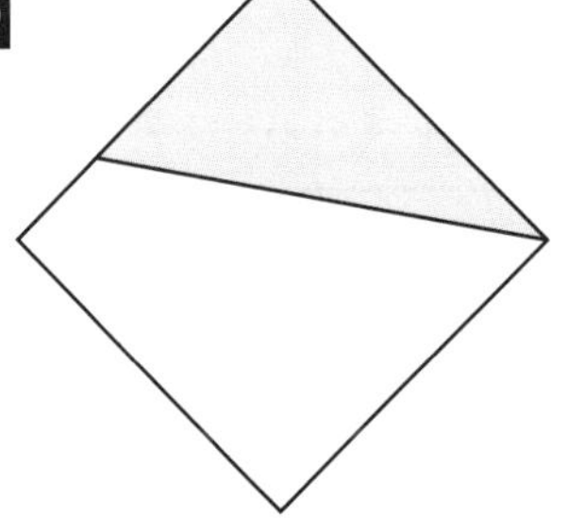

14
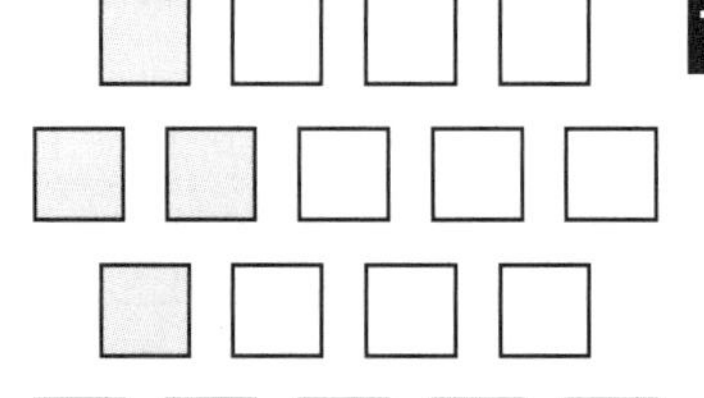

15
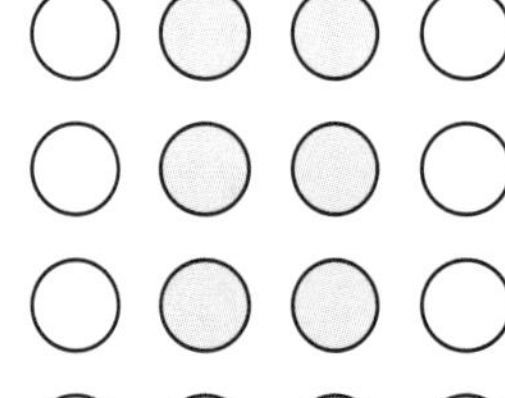

Recognise quarters of shapes and groups

Which of these shapes or groups show one quarter ($\frac{1}{4}$)?

1

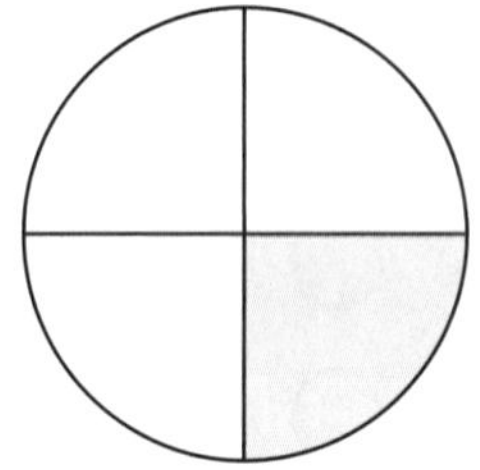

2

3

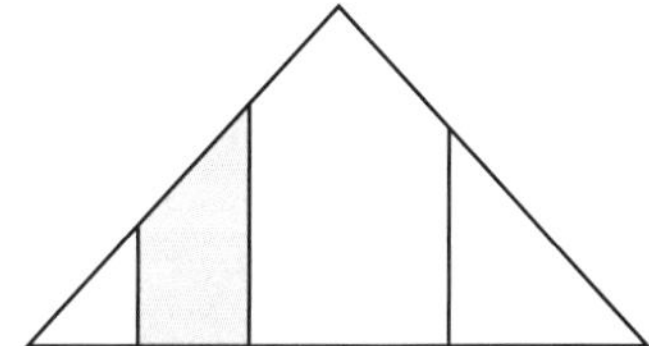

4

5

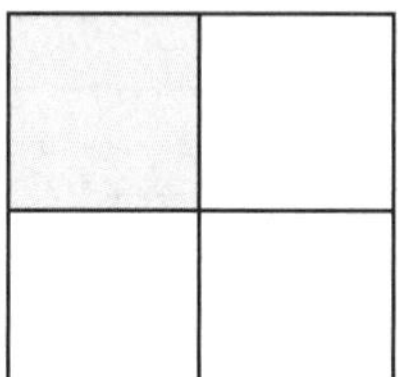

6

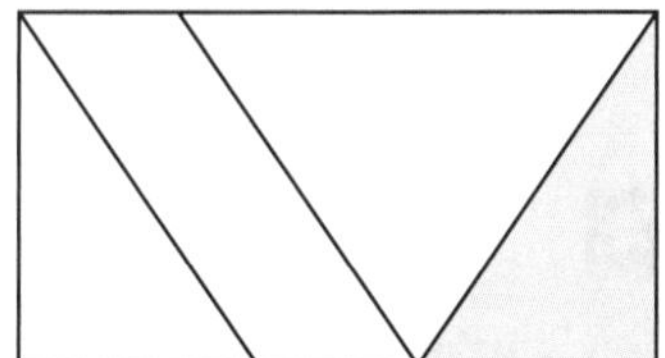

7

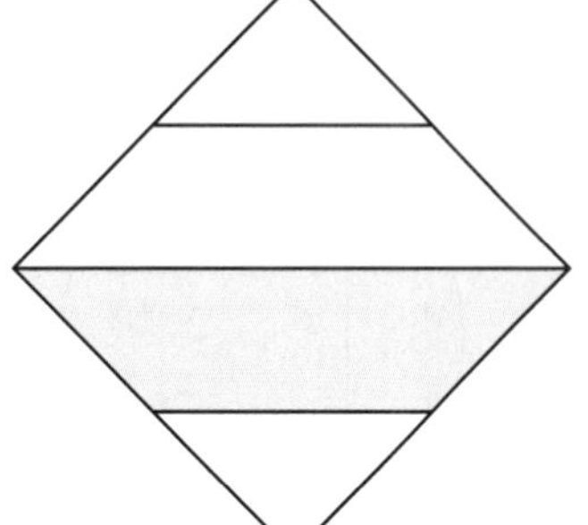

8

9

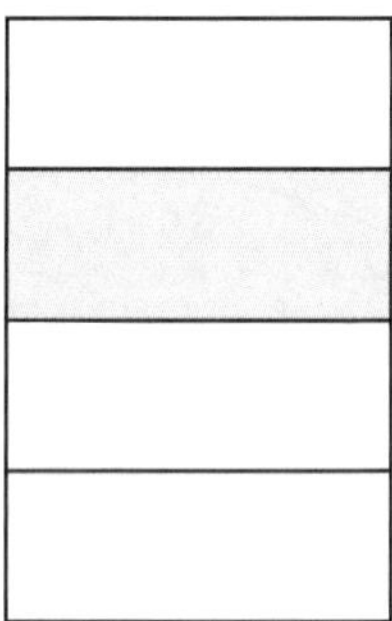

10

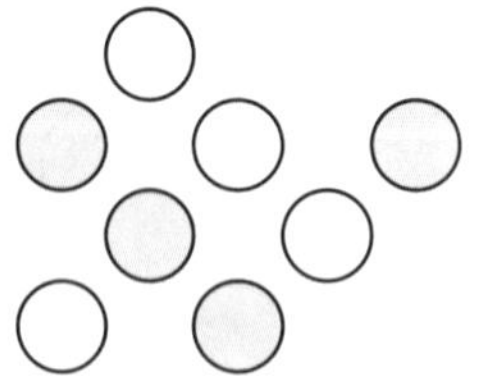

11

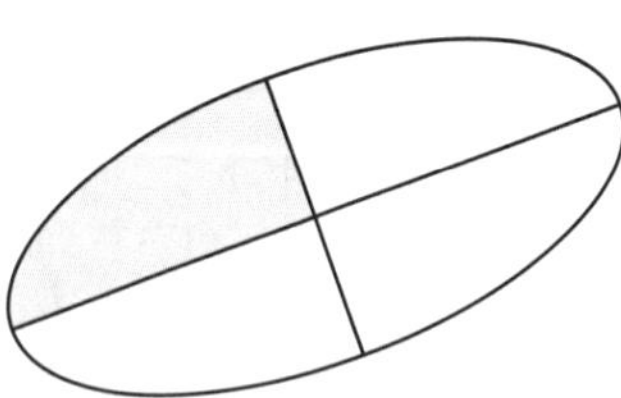

12

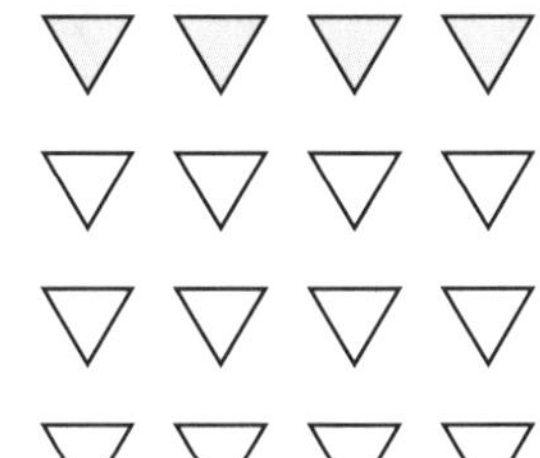

13

14

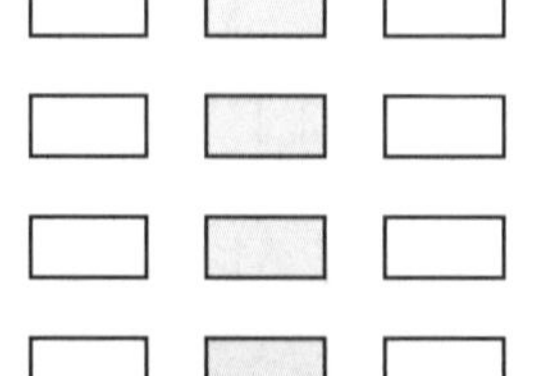

15

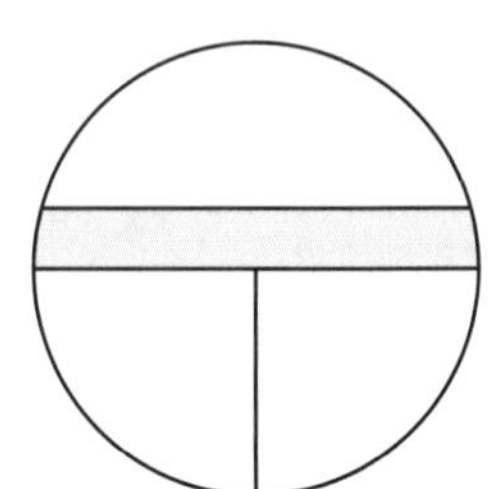

Recognise thirds of shapes and groups

Which of these shapes or groups show one third ($\frac{1}{3}$)?

1

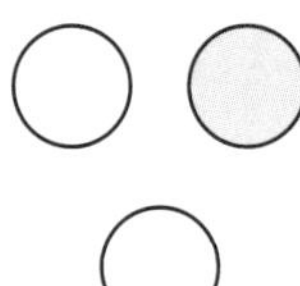

2

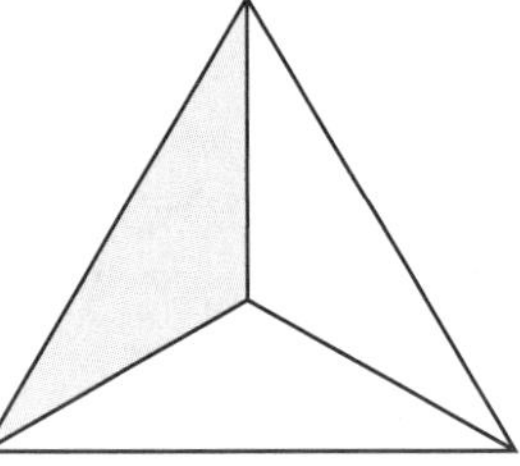

3

4

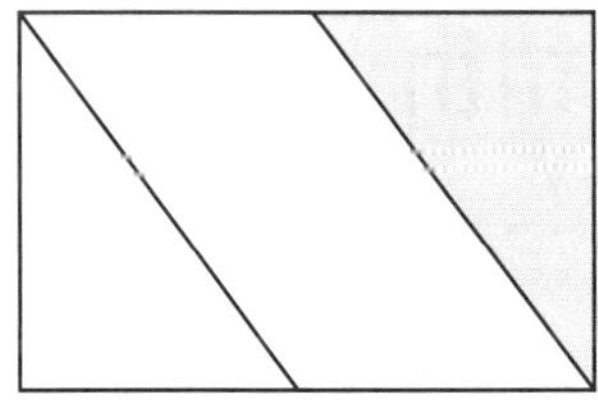

5

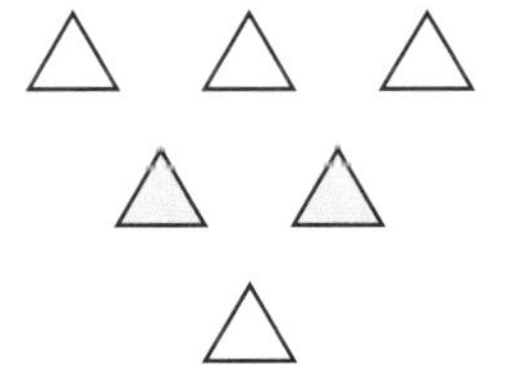

6

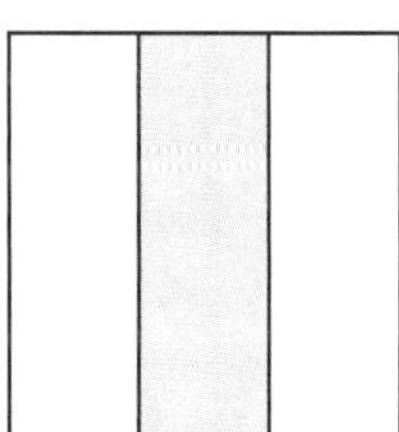

7

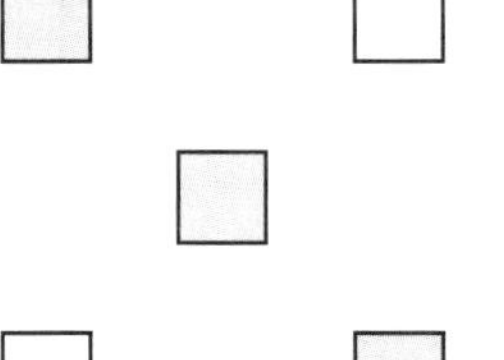

8

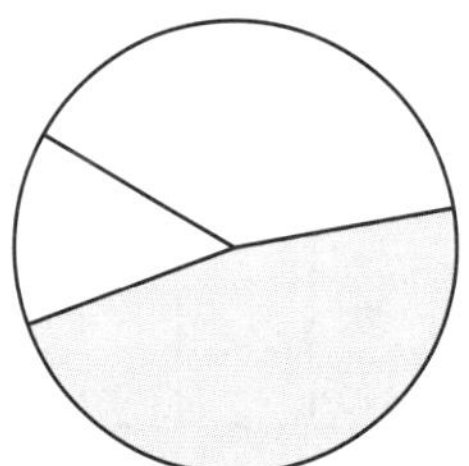

9

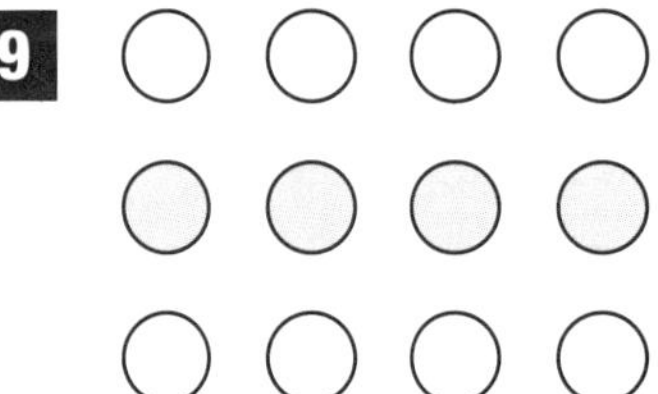

10

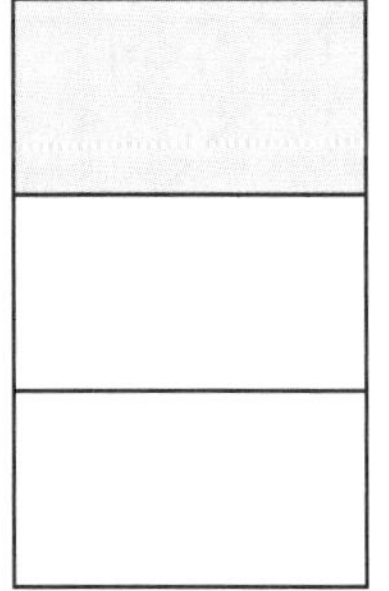

11

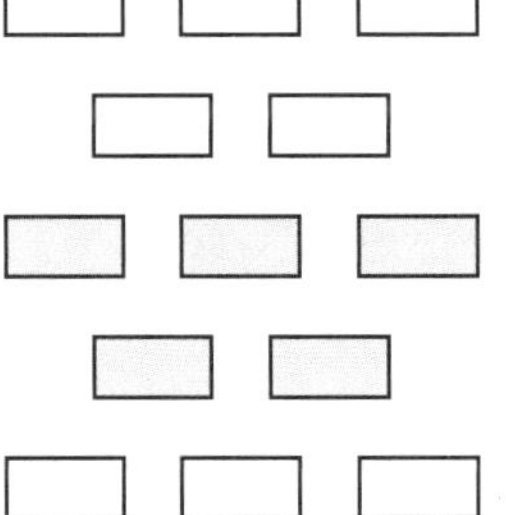

12

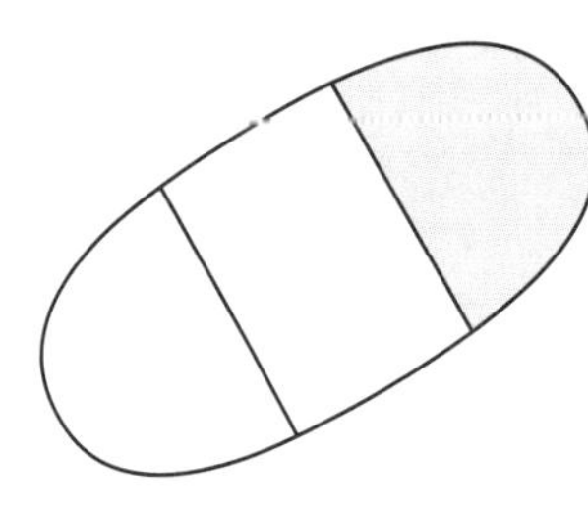

13

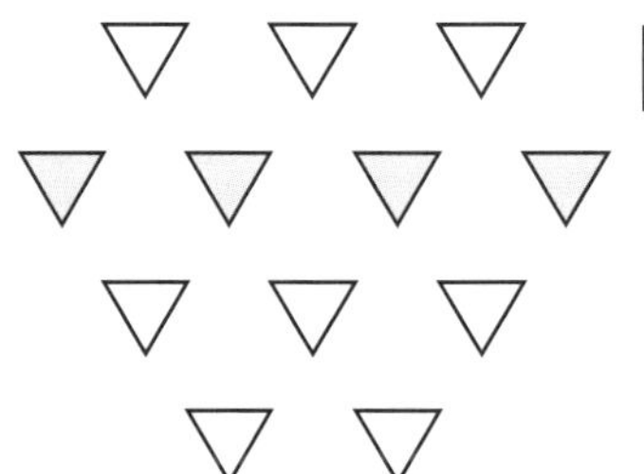

14

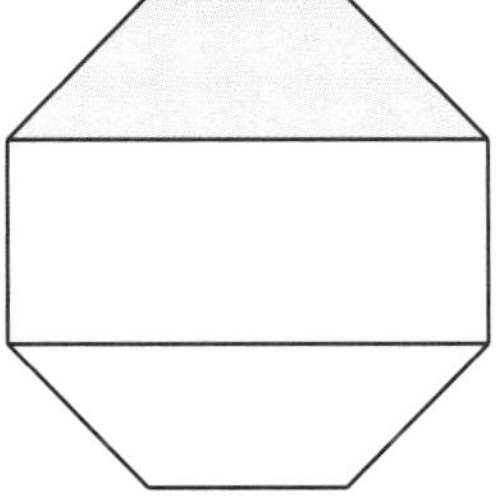

15

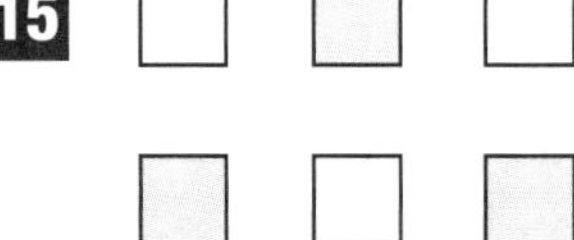

Match fraction symbols with diagrams and words

Write each fraction symbol and its matching words, and then write the letter of all the diagrams that match it.

1 $\frac{1}{2}$

2 $\frac{1}{4}$

3 $\frac{1}{3}$

4 $\frac{1}{5}$

5 $\frac{1}{10}$

A

B

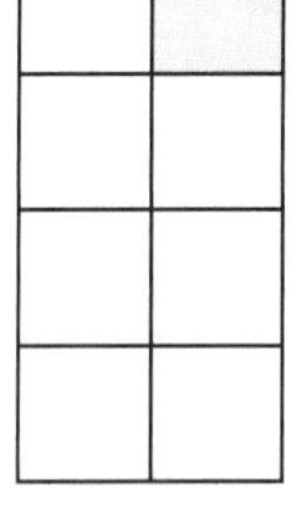

C

one fifth

one third

D

E

one tenth

F

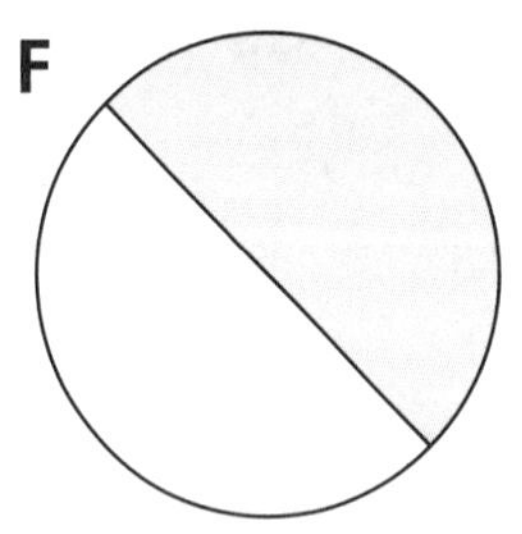

one quarter

H

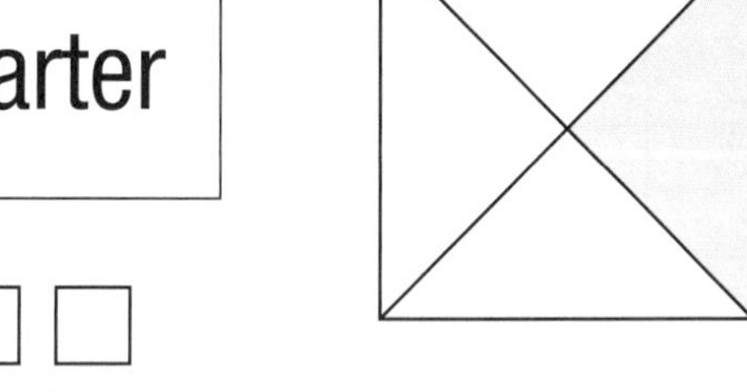

G

I

J

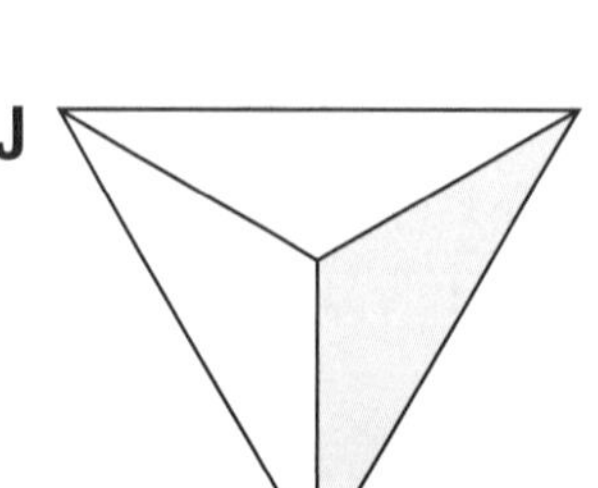

one half

Match fraction symbols with diagrams

Write each fraction symbol, and then write the letter of each diagram that matches it.

1

A

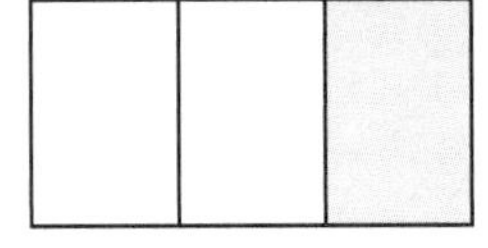

B

C

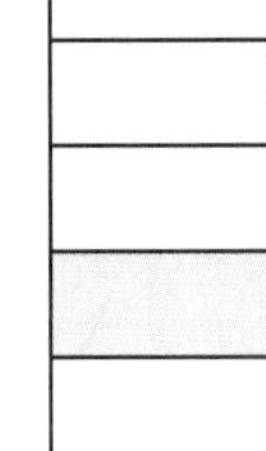

E

D

2

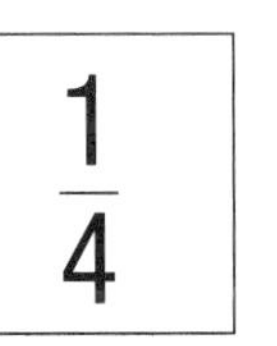

F

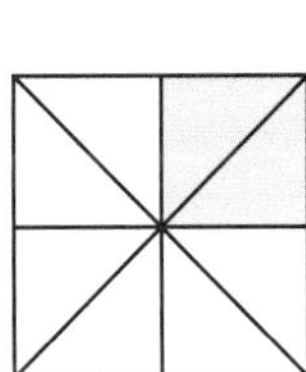

G

H

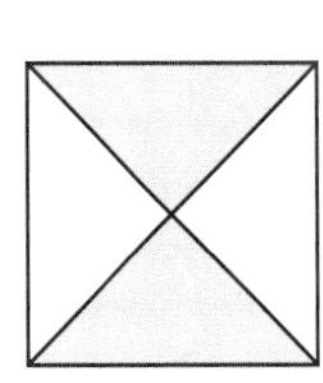

I

3

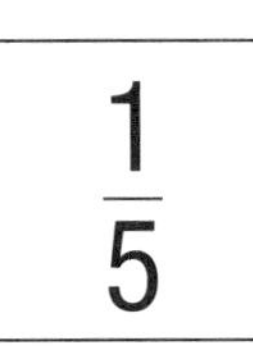

J

K

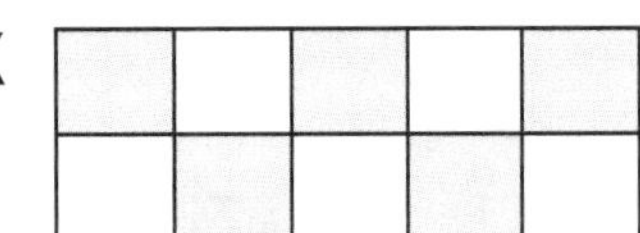

M

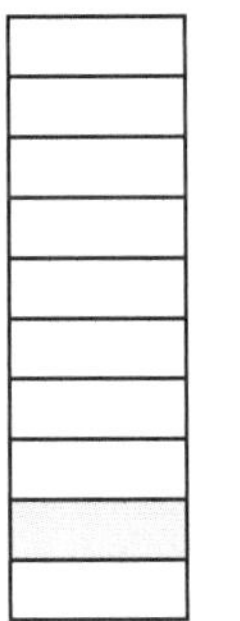

N

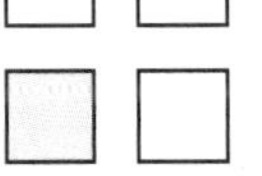

L

4

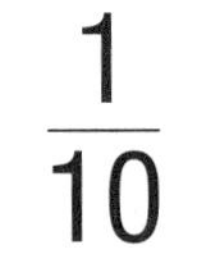

O

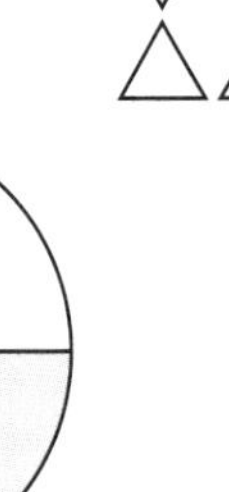

5

P

Q

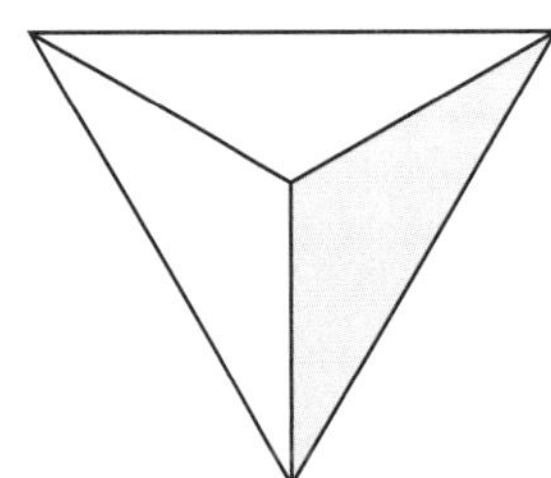

Represent fractions of shapes and groups

What fraction of each shape or group has been shaded? Write the fractions in symbols and words.

1

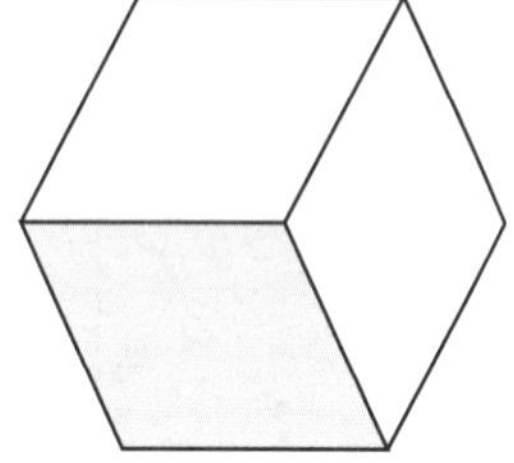

2

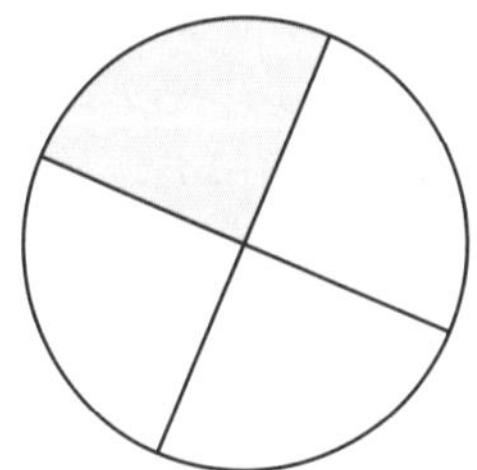

3

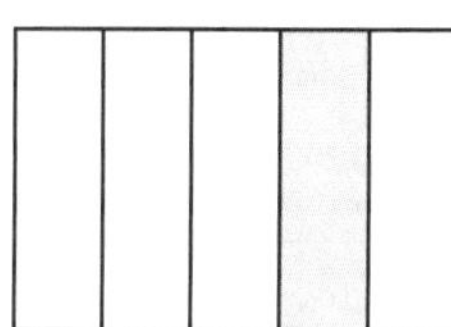

4

5

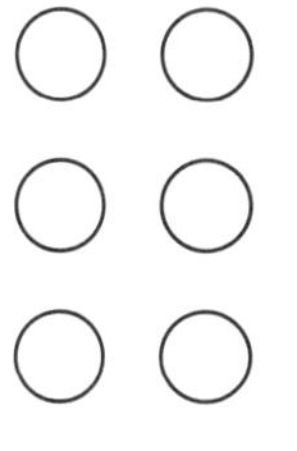

6

7

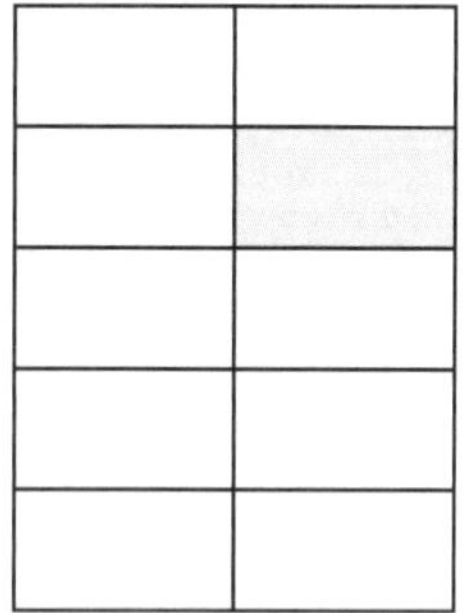

8

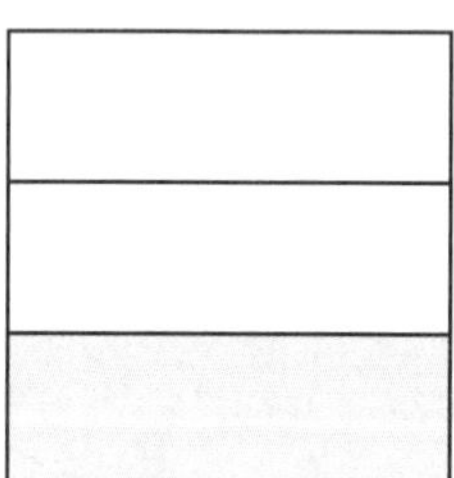

9

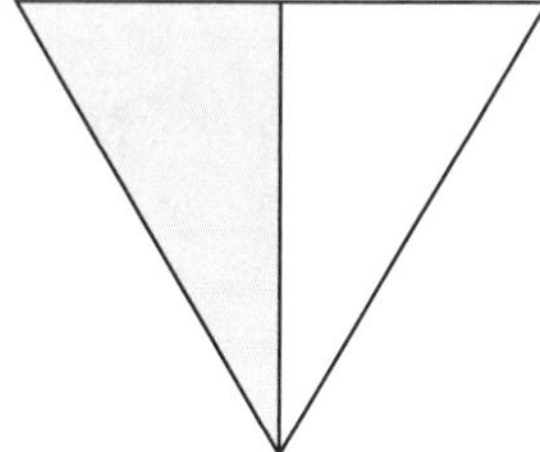

10

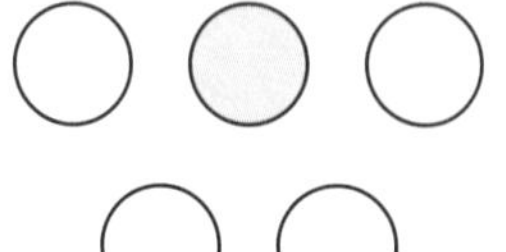

11

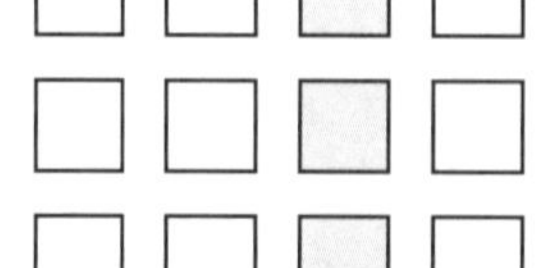

12

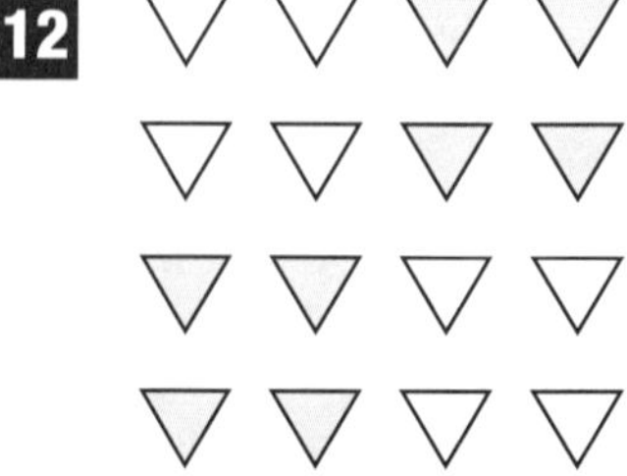

Identify decimal notation in real life situations

Match coins with symbols

Which money jar in each row has the amount of money shown on the left?

1 K1.70

a 

b 

c 

2 K1.25

a 

b 

c 

3 K2.50

a 

b 

c 

4 K2.00

a 

b 

c 

5 K1.95

a 

b 

c 

Match coins and notes with money symbols

1 Write the notes and coins that make the amounts of money on the left. If possible, do it in more than one way.

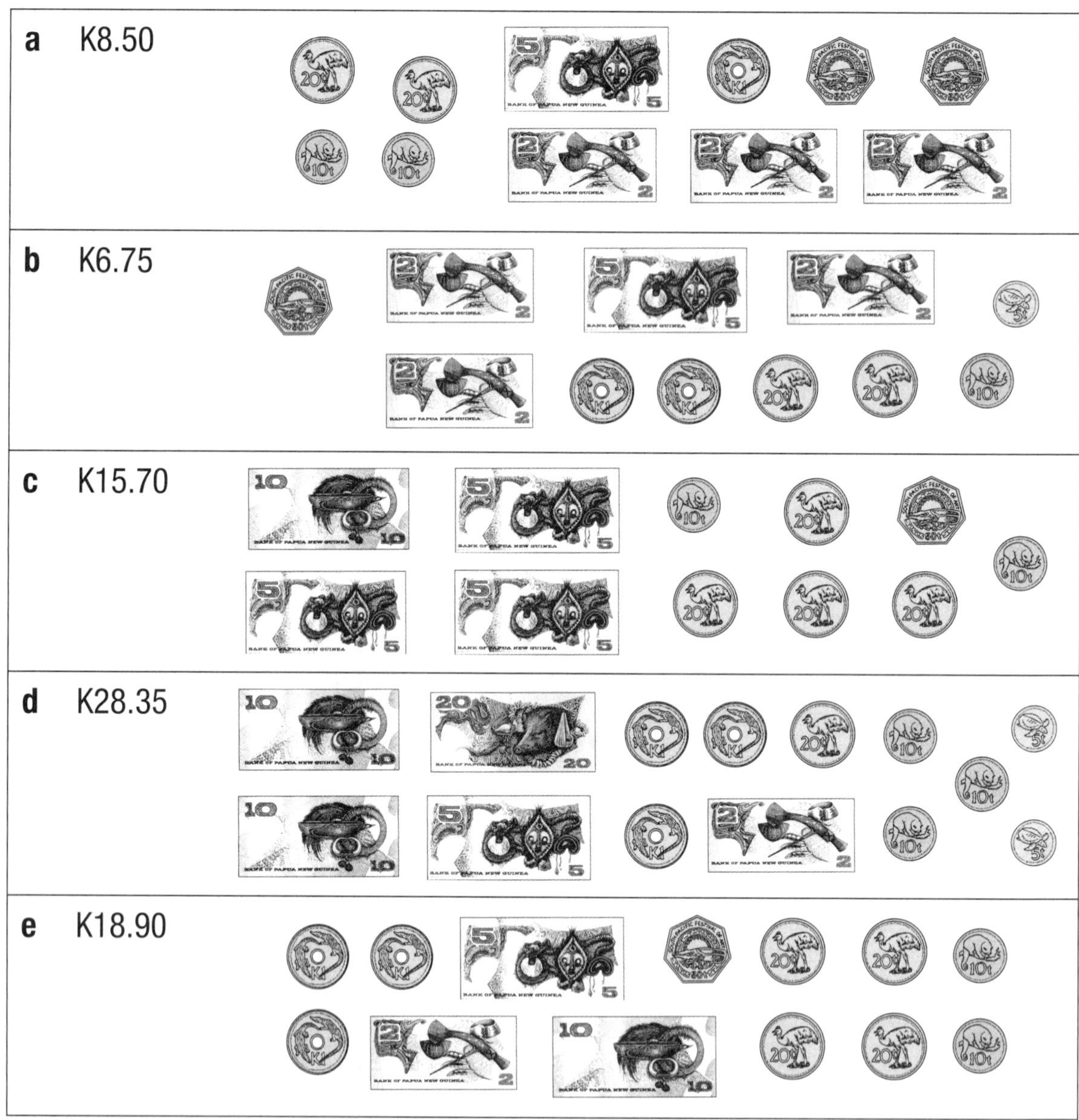

a	K8.50
b	K6.75
c	K15.70
d	K28.35
e	K18.90

2 Write the notes and coins that you would choose to pay for each item.

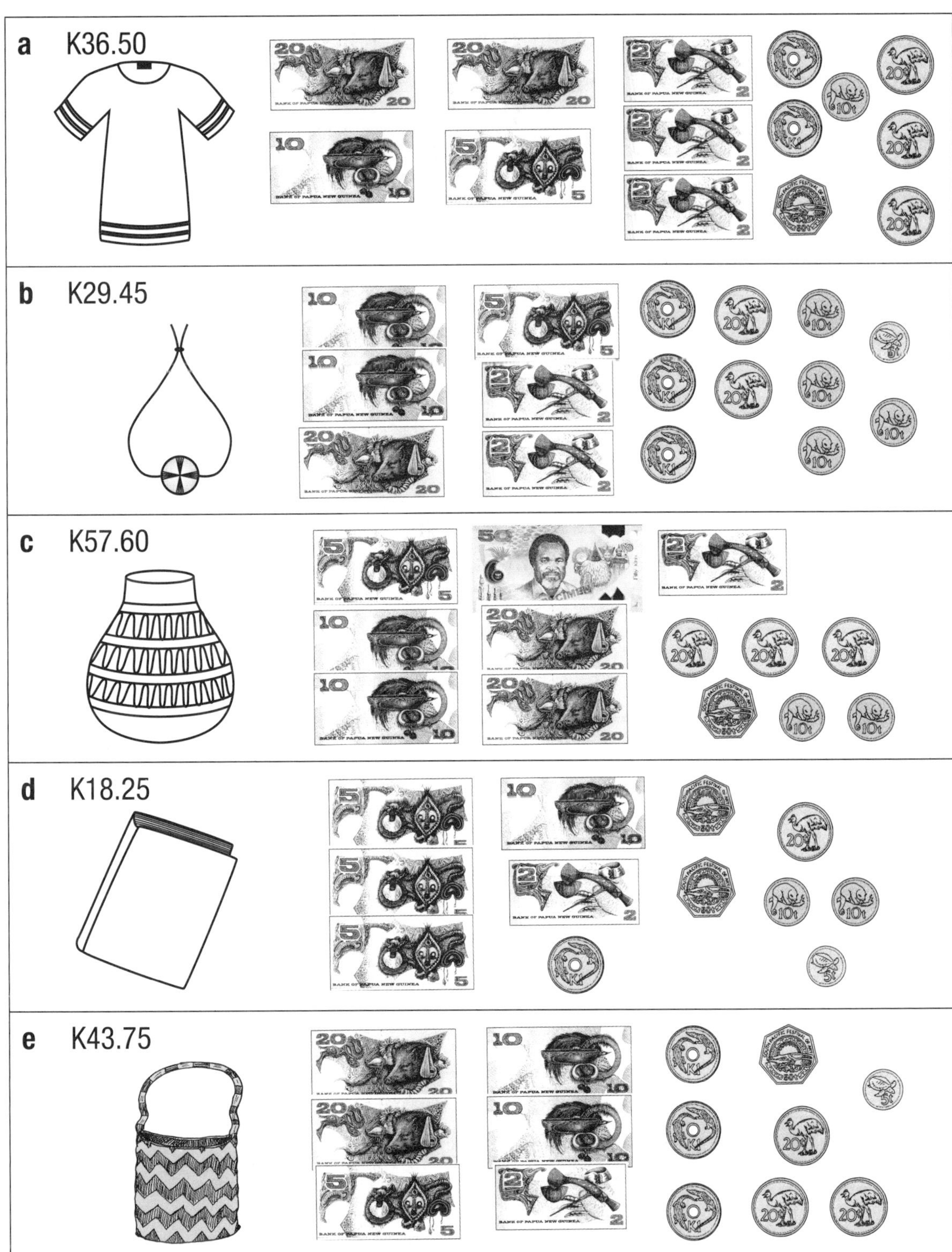

Make amounts of money

Draw the coins or notes that you need to add to each frame to make the amount shown and then write how much you added.

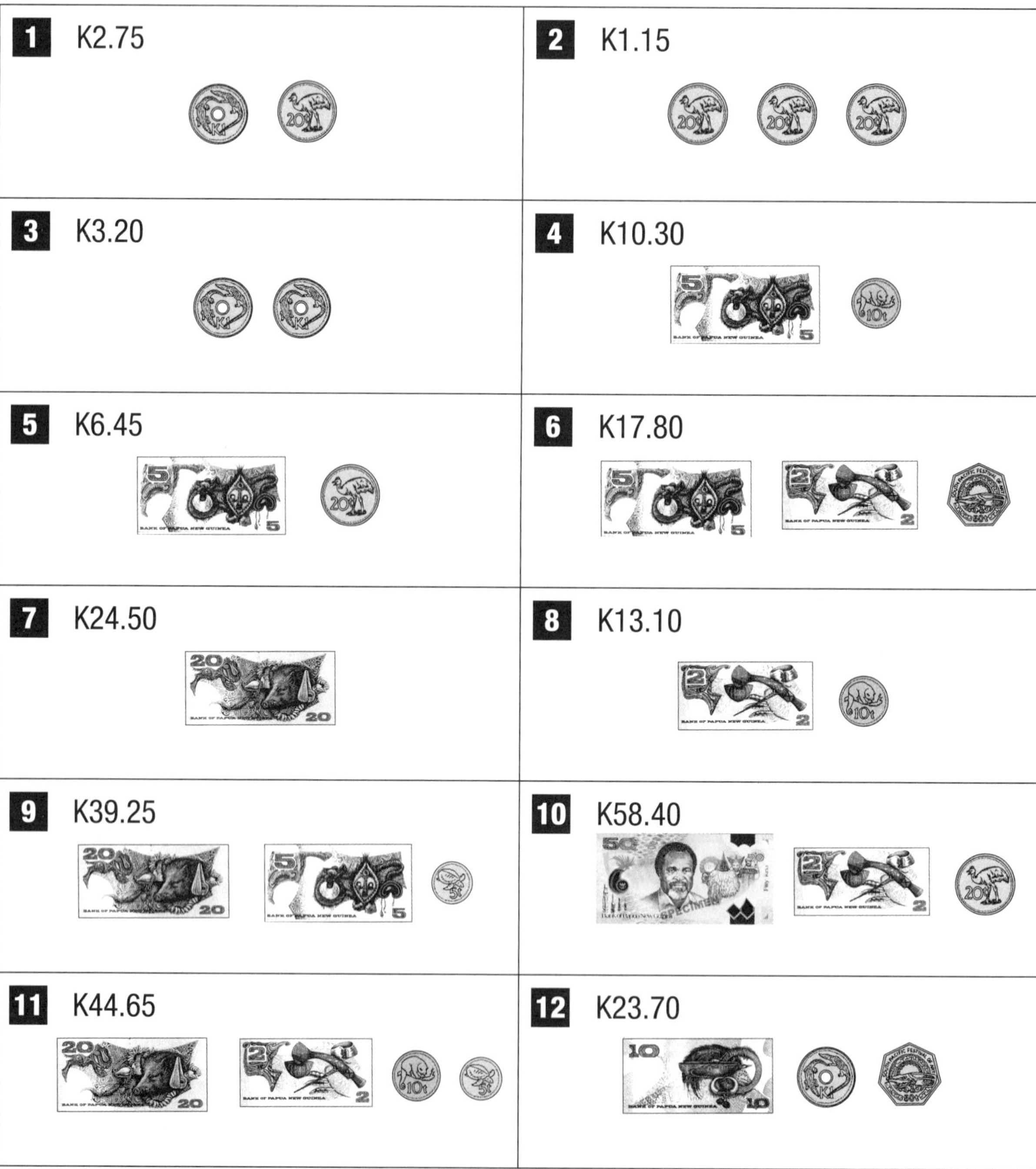

1. K2.75
2. K1.15
3. K3.20
4. K10.30
5. K6.45
6. K17.80
7. K24.50
8. K13.10
9. K39.25
10. K58.40
11. K44.65
12. K23.70

Count and record money

1 How much money is in each purse?

a

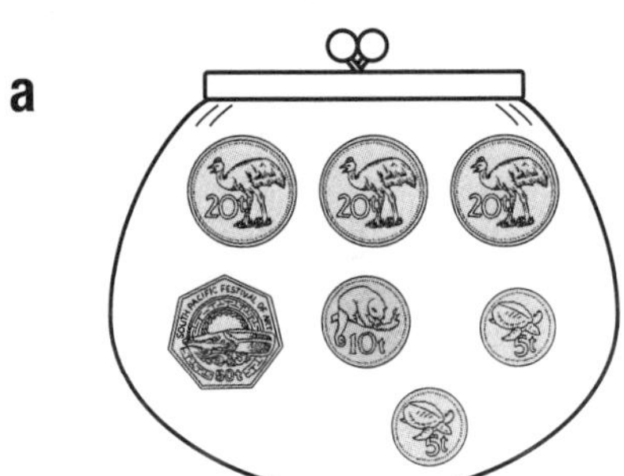

b

c

d

2 How much money is in each wallet?

a

b

c

d

3 How much money is in each money box?

4 Which money box has the most money?

5 Which money box has the least money?

6 Put string around groups of coins that make K1, and then work out how much money is on the page.

Give change from K1

For each item, draw the coins you would give as change from K1.

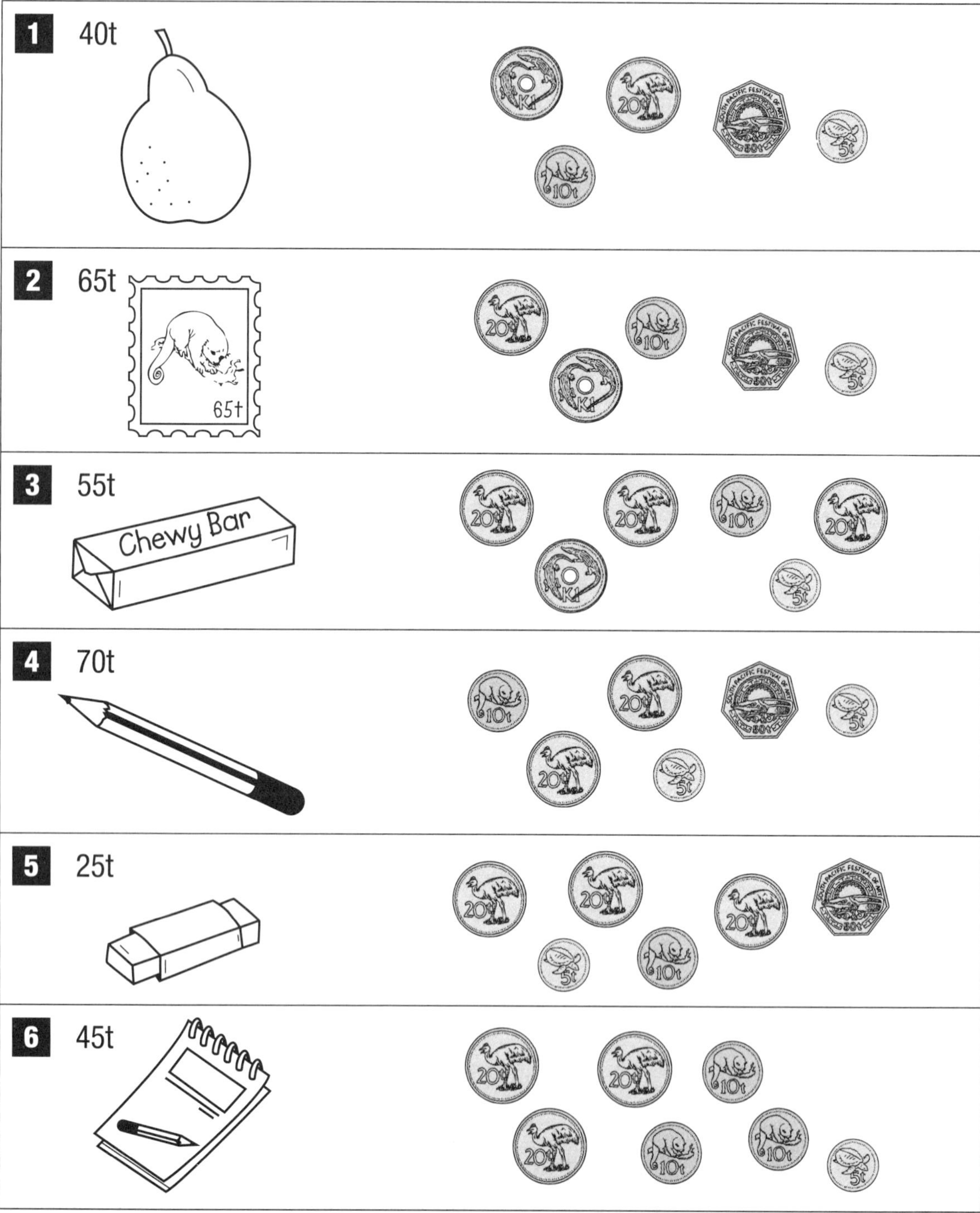

Give change from K2

For each item, draw the coins you would give as change from K2.

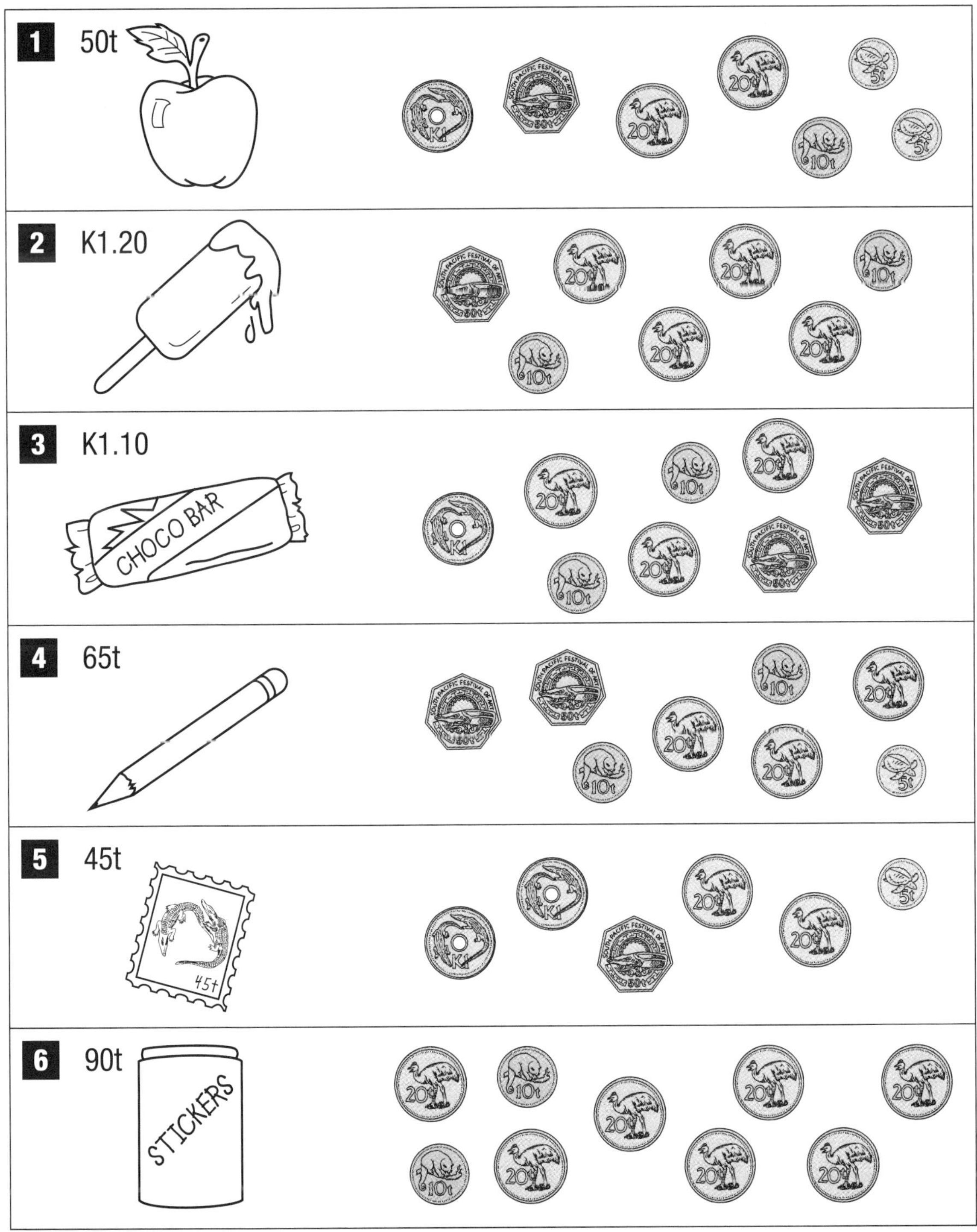

Calculate with money

1 Add these amounts.

- **a** K2.60 + K0.50
- **b** K1.35 + K1.90
- **c** K3.50 + K2.75
- **d** K0.65 + K2.55
- **e** K1.80 + K3.30
- **f** K5.00 + K2.45 + K1.70
- **g** K4.10 + K2.95 + K3.00
- **h** K3.15 + K2.00 + K4.90
- **i** K2.15 + K5.60 + K1.25

2 Subtract these amounts.

- **a** K5.00 – K2.50
- **b** K4.20 – K1.65
- **c** K3.15 – K1.50
- **d** K10.00 – K5.45
- **e** K2.55 – K0.75
- **f** K8.35 – K4.70
- **g** K5.10 – K2.85
- **h** K7.40 – K3.60
- **i** K8.00 – K5.15

3 Multiply these amounts.

- **a** K2.50 × 2
- **b** K0.80 × 3
- **c** K1.60 × 2
- **d** K4.30 × 4
- **e** K3.20 × 5
- **f** K2.15 × 3
- **g** K4.50 × 6
- **h** K7.50 × 3

4 Divide these amounts.

- **a** K6.30 ÷ 3
- **b** K8.60 ÷ 2
- **c** K9.00 ÷ 3
- **d** K4.80 ÷ 4
- **e** K10.80 ÷ 2
- **f** K7.50 ÷ 3
- **g** K12.50 ÷ 5
- **h** K5.20 ÷ 4

Match money word problems with number sentences

Write the money number sentence that matches each word problem.

1 Kennedy buys 3 bags of vegetables at K8.35 per bag. What is the total cost of the 3 bags?

2 At the market, Omah buys goods worth K12.85. If he gives K20 to the stall owner, how much change should he get?

K30.40 ÷ 4 = K7.60

K35.50 ÷ 5 = K7.10

4 Grace earned K35.50 and K31.70. How much did she earn in total?

3 Ella bought a book at K9.50 and a pen at K2.65. How much did she spend?

K20.00 – K12.85 = K7.15

K9.50 × 5 = K47.50

5 Ben makes K9.50 per day at a part-time job. How much does he make after 5 days?

6 Five villagers shared equally the cost of garden equipment. If the total cost was K35.50, how much did each villager pay?

K8.35 × 3 = K25.05

K12.85 – K8.35 = K4.50

7 One shirt costs K12.85 and another shirt costs K8.35. What is the difference in price between the two shirts?

8 Four books cost K30.40 in total. If each book costs the same, how much does one book cost?

K9.50 + K2.65 = K12.15

K35.50 + K31.70 = K67.20

Multiple choice test for Number and place value (3.1.1)

1 Which number matches the words *ninety-seven*?

a 907 **b** 97

c 9007 **d** 970

2 Which number matches the words *two hundred and fifty-six*?

a 20 056 **b** 2056

c 256 **d** 2506

3 Which number comes one after *one hundred and seventy-nine*?

a 178 **b** 180

c 1800 **d** 170

4 If there are five people in front of you, what position are you in the line?

a 5th **b** 7th

c 1st **d** 6th

5 Which number is largest?

a 54 **b** 45

c 58 **d** 50

6 Which number is smallest?

a 857 **b** 875

c 758 **d** 785

7 Which number has 3 tens?

a 113 **b** 237

c 309 **d** 43

8 Which number has 9 ones?

a 96 **b** 195

c 29 **d** 985

9 Which number has 0 tens?

a 150 b 205

c 290 d 110

10 What is the largest number you can make from 2, 7 and 3?

a 732 b 327

c 237 d 723

11 Which number is closest to 500?

a 560 b 420

c 519 d 495

12 What number is: 10, 9 and 400?

a 419 b 491

c 400 109 d 40 019

13 Which of these numbers will you say when counting by 5s starting at 83?

a 100 b 113

c 105 d 90

14 What is the next number? 69, 79, 89, 99, ______

a 100 b 19

c 1009 d 109

15 What is the next number? 234, 224, 214, 204, ______

a 203 b 200

c 194 d 184

16 Which number is even?

a 518 b 139

c 223 d 605

17 Which of these additions has an answer that is even?

a 22 + 7 b 17 + 15

c 19 + 14 d 23 + 12

Assessment Number and Application

Operations (3.1.2)

1 Add each set of numbers in your head and write the answers.

a 7 + 5 + 3 + 4 + 5 **b** 6 + 7 + 4 + 7 + 3

c 4 + 8 + 6 + 2 + 8 **d** 9 + 5 + 1 + 9 + 5

2 Solve these number sentences in your head and write the answers.

a 40 + 50 **b** 70 – 20 **c** 80 – 50 **d** 60 + 50

e 90 – 40 **f** 30 + 50 **g** 70 + 60 **h** 80 – 30

3 Add or subtract 10 in your head and write the answers.

a 74 – 10 **b** 96 + 10 **c** 143 – 10 **d** 127 + 10

e 212 – 10 **f** 105 + 10 **g** 271 – 10 **h** 349 + 10

4 Use doubles to help you answer these additions.

a 24 + 25 **b** 18 + 17 **c** 41 + 42 **d** 15 + 16

e 37 + 36 **f** 21 + 22 **g** 55 + 54 **h** 27 + 28

5 Add or subtract 9 or 11 in your head and write the answers.

a $28 + 9$ **b** $56 - 9$ **c** $34 + 11$ **d** $85 - 11$

e $127 + 11$ **f** $152 - 11$ **g** $209 + 9$ **h** $243 - 9$

6 Add or subtract 99 in your head and write the answers.

a $55 + 99$ **b** $372 - 99$ **c** $217 + 99$ **d** $161 - 99$

7 How many more are needed to make 100?

a 45 **b** 73 **c** 38 **d** 61 **e** 54 **f** 27

8 How many more are needed to make 1000?

a 600 **b** 350 **c** 270 **d** 410 **e** 780 **f** 695

9 Copy and complete each addition or subtraction.

a $18 - \square = 9$ **b** $\square + 7 = 15$ **c** $17 - \square = 8$ **d** $9 + \square = 16$

e $\square - 6 = 12$ **f** $6 + \square = 14$ **g** $19 - \square = 7$ **h** $\square + 8 = 17$

10 Multiply these numbers in your head and write the answers.

a 8×2 **b** 2×12 **c** 5×6 **d** 9×5 **e** 3×10 **f** 10×8

g 11×5 **h** 3×5 **i** 10×6 **j** 12×10 **k** 7×2 **l** 2×9

11 Divide these numbers in your head and write the answers.

a	$12 \div 2$	**b**	$20 \div 5$	**c**	$40 \div 10$	**d**	$16 \div 2$
e	$100 \div 10$	**f**	$15 \div 5$	**g**	$70 \div 10$	**h**	$24 \div 2$
i	$30 \div 5$	**j**	$120 \div 10$	**k**	$25 \div 5$	**l**	$14 \div 2$

12 Solve these additions.

a $\begin{array}{r} 54 \\ +35 \\ \hline \end{array}$ **b** $\begin{array}{r} 63 \\ +54 \\ \hline \end{array}$ **c** $\begin{array}{r} 48 \\ +76 \\ \hline \end{array}$ **d** $\begin{array}{r} 37 \\ +85 \\ \hline \end{array}$

e $\begin{array}{r} 235 \\ +342 \\ \hline \end{array}$ **f** $\begin{array}{r} 475 \\ +366 \\ \hline \end{array}$ **g** $\begin{array}{r} 247 \\ +584 \\ \hline \end{array}$ **h** $\begin{array}{r} 364 \\ +\ \ 77 \\ \hline \end{array}$

13 Solve these subtractions.

a $\begin{array}{r} 67 \\ -24 \\ \hline \end{array}$ **b** $\begin{array}{r} 78 \\ -45 \\ \hline \end{array}$ **c** $\begin{array}{r} 87 \\ -38 \\ \hline \end{array}$ **d** $\begin{array}{r} 92 \\ -56 \\ \hline \end{array}$

e $\begin{array}{r} 248 \\ -127 \\ \hline \end{array}$ **f** $\begin{array}{r} 416 \\ -265 \\ \hline \end{array}$ **g** $\begin{array}{r} 524 \\ -279 \\ \hline \end{array}$ **h** $\begin{array}{r} 731 \\ -\ \ 68 \\ \hline \end{array}$

14 Solve these multiplications.

a $\begin{array}{r} 42 \\ \times 3 \\ \hline \end{array}$ **b** $\begin{array}{r} 31 \\ \times 4 \\ \hline \end{array}$ **c** $\begin{array}{r} 37 \\ \times 5 \\ \hline \end{array}$ **d** $\begin{array}{r} 62 \\ \times 8 \\ \hline \end{array}$

15 Complete the gaps in these division number sentences.

a $23 \div 5 = \square$ rem. $\square$

b $31 \div 10 = \square$ rem. $\square$

c $37 \div 5 = \square$ rem. $\square$

d $19 \div 2 = \square$ rem. $\square$

16 Solve these word problems.

a A group of people has travelled 158 kilometres out of a total distance of 325 kilometres. How much further does the group have to travel?

b Kapi has paid K147 for repairs to his car. He still has to pay another K85. What is the total amount Kapi has to pay?

c Joe shared K60 equally among his 5 children. How much did each child get?

d Each container holds 28 litres of water. If there are 6 containers, how many litres of water are there altogether?

e Esta planted 215 seedlings. Naomi planted 166 seedlings. How many more seedlings did Esta plant?

Assessment — Number and Application

Fractions (3.1.3)

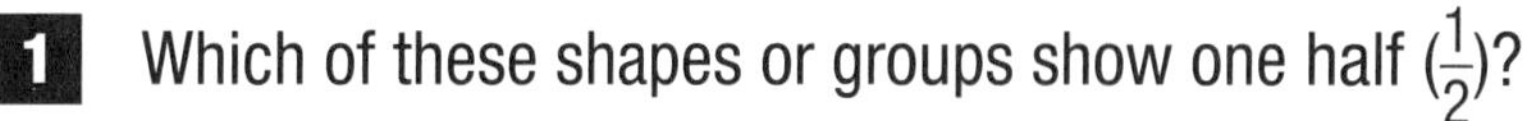

1 Which of these shapes or groups show one half ($\frac{1}{2}$)?

a 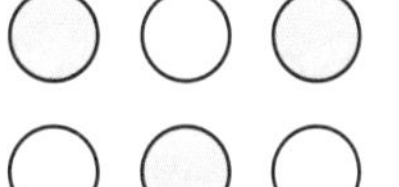

b 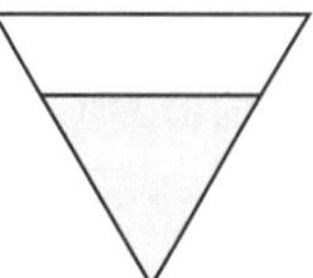

c 

d 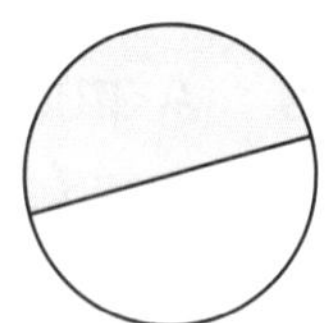

2 Which of these shapes or groups show one quarter ($\frac{1}{4}$)?

a 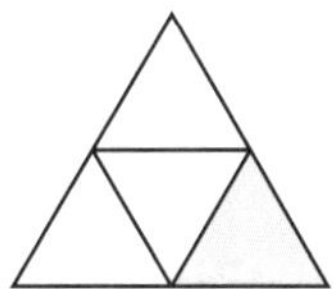

b 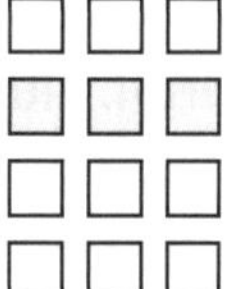

c 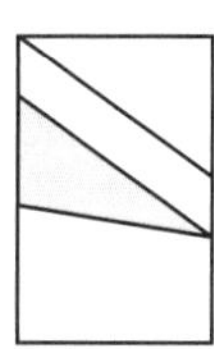

d

3 Which of these shapes or groups show one third ($\frac{1}{3}$)?

a

b 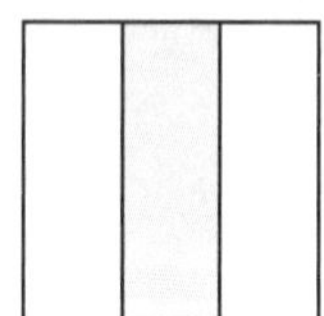

c 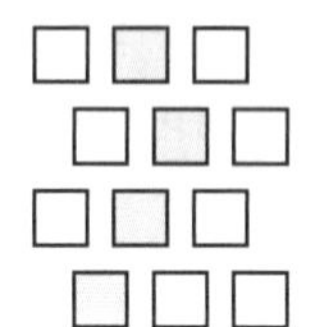

d 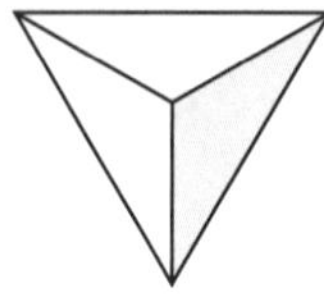

4 What fraction of each shape or group has been shaded?

a 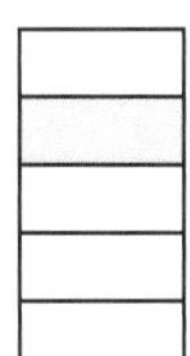

b 

c 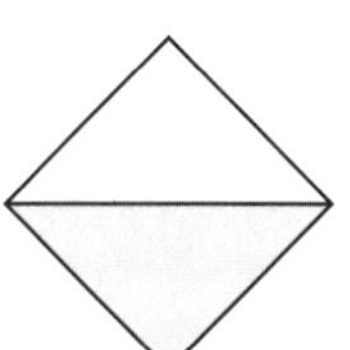

d 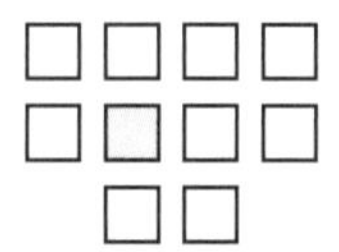

e

f 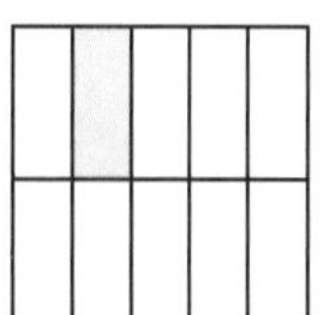

g 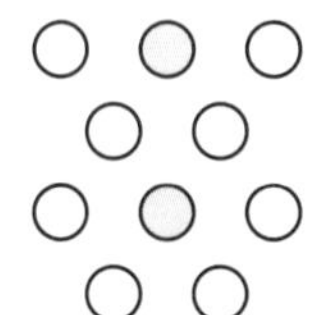

h 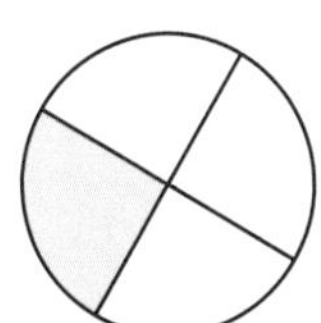

Decimals (3.1.4)

1 Write or draw two different ways that each amount of money can be made.

a K7.50 **b** K6.75 **c** K14.30

2 Write or draw the notes and coins that you would give for these amounts. Do this using the least possible number of notes and coins.

a K12.60 **b** K28.45 **c** K17.80

3 Write or draw the least number of notes and coins that you would add to each box to make the amount shown.

a K32.20

b K49.25

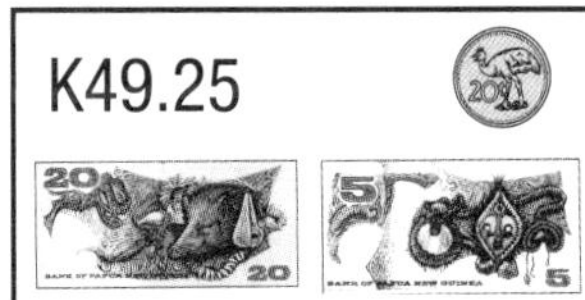

c K26.85

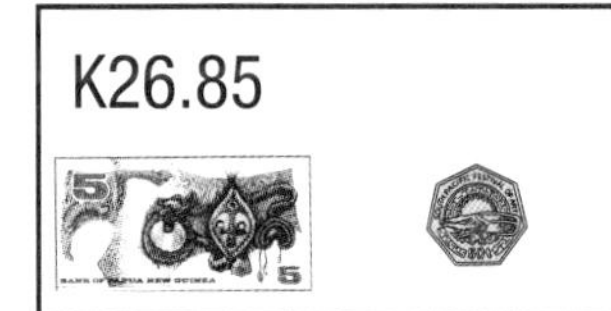

4 How much money is in each money box?

a

b

c

5 Do these money calculations.

a K1.80 + K0.70 **b** K5.20 – K2.70 **c** K3.25 + K2.90

d K1.70 × 2 **e** K4.20 ÷ 4 **f** K7.35 – K 4.50

Strand	Measurement

Estimate and measure lengths, distances and perimeters using formal and informal units

Estimate and measure length in centimetres

1 Draw a chart like the one shown. Estimate the length of each line in centimetres (cm) and record it in the chart. Then measure each line and record its actual length in the chart.

Estimate	Actual length

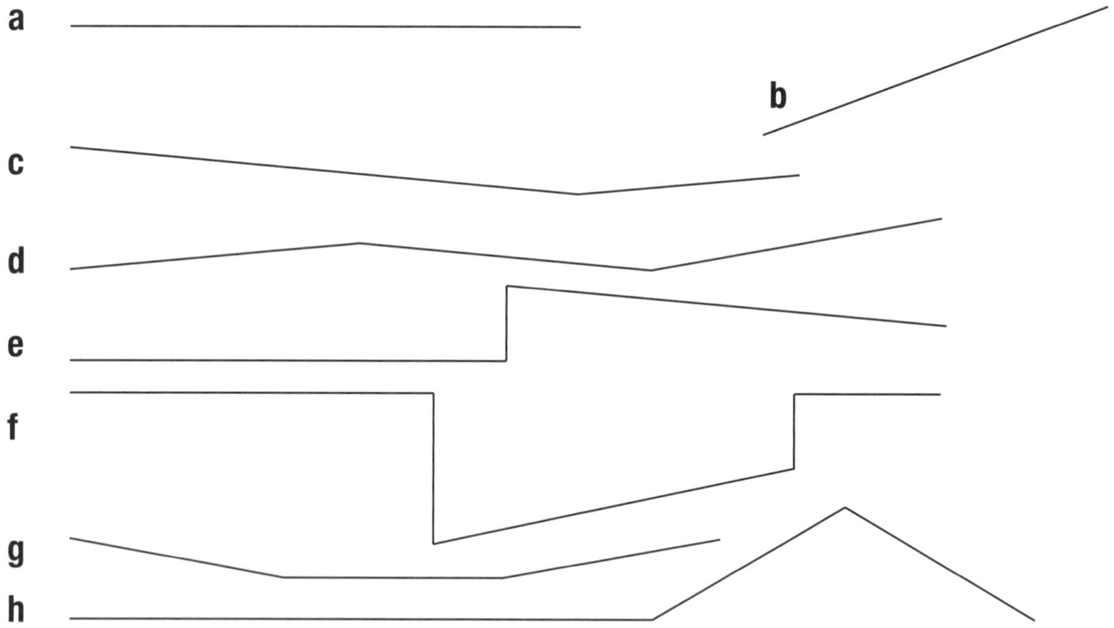

a

b

c

d

e

f

g

h

2 Which line is longest?

3 How much longer than line C is line F?

4 What is the difference in length between:

a line A and line D? **b** line B and line G? **c** line E and line H?

Measure perimeter in centimetres

1 Measure the perimeter of each shape in centimetres.

a 

b 

c

d 

e 

2 Which shape has the longest perimeter?

3 Guess which shape has the longest perimeter: a, b, c or d?

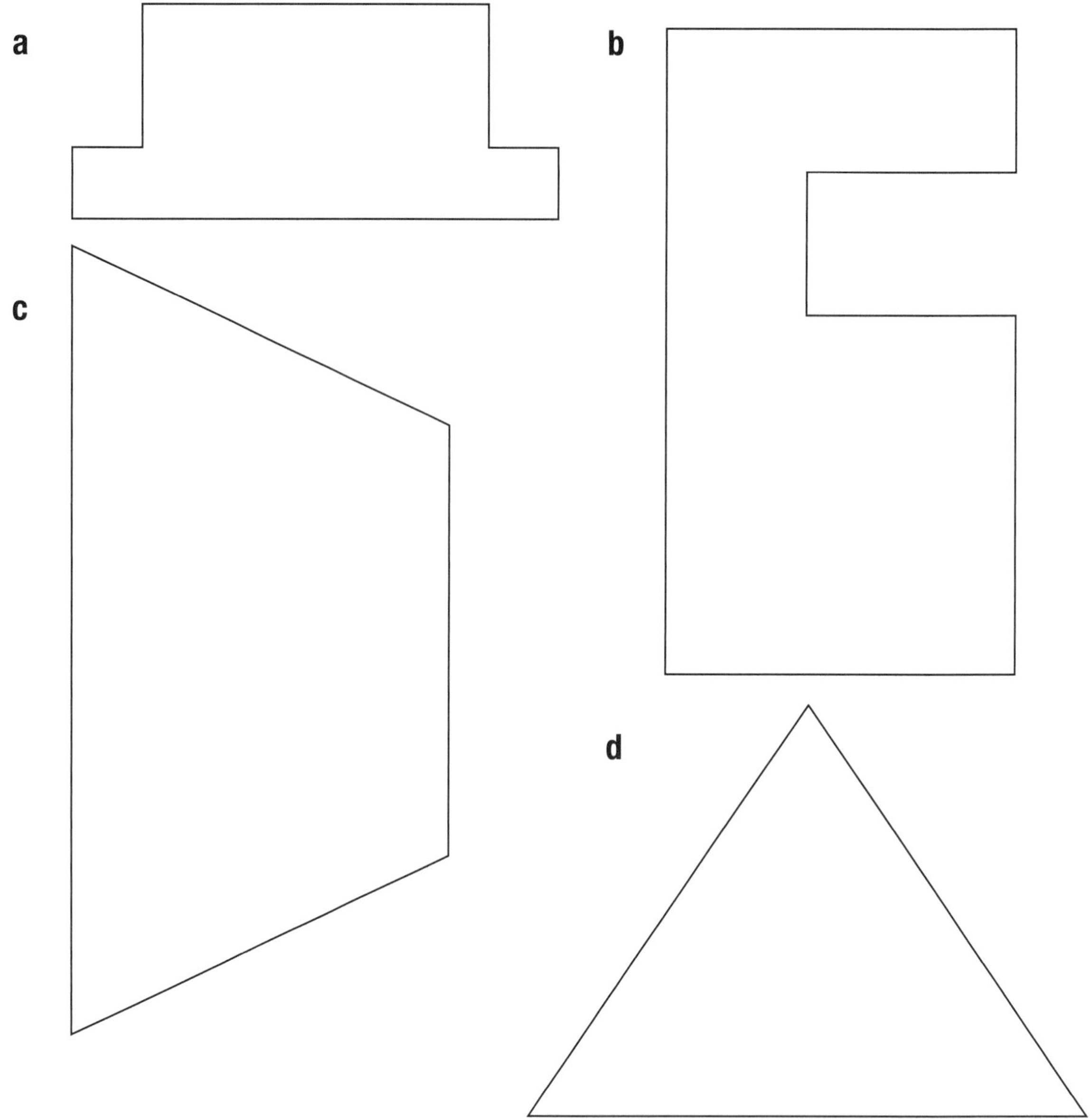

4 Guess which shape has the shortest perimeter: a, b, c or d?

5 Measure each perimeter and write it in your book. Were your guesses correct?

Estimate and measure areas using informal regular units

Count square units to find area

Count the square units to find the area of each shape. The first shape has an area of 7 square units.

1

2

3

4

5

Count square units to compare areas

1 Count the square units in each shape and record the area in square units.

a

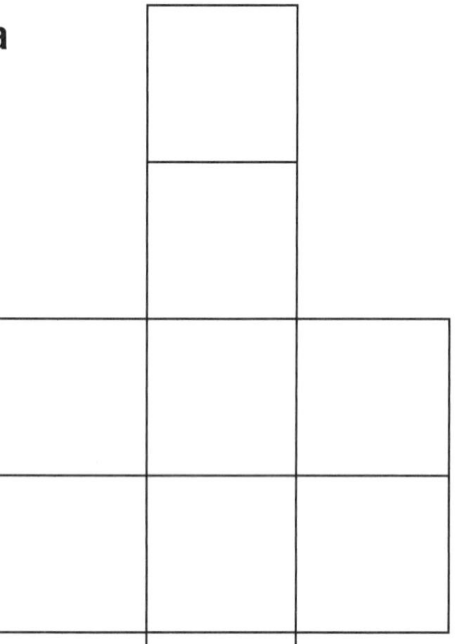

b

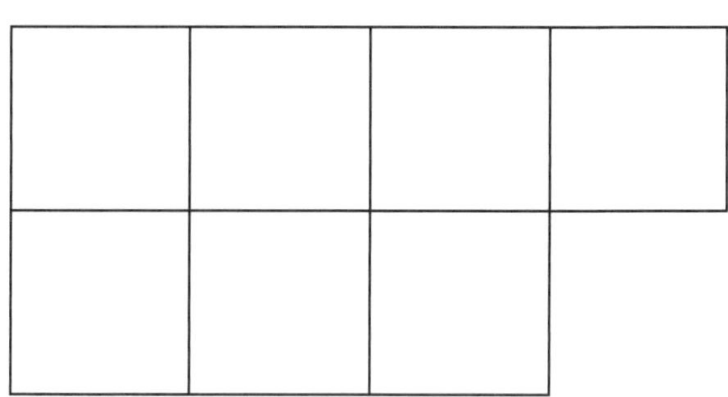

c

d

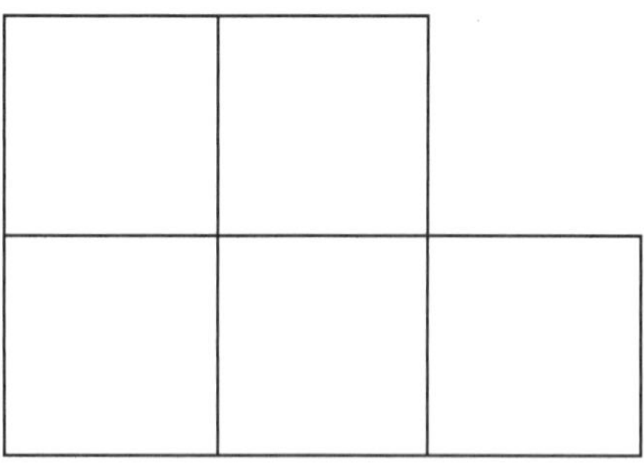

e

2 Which shape has the largest area?

3 Which has the smallest area?

4 How many square units larger is:

a C than E? **b** A than D? **c** B than E? **d** C than D?

Estimate and measure volume and capacity using informal units

Count cubes to find volume

1 Count the cubes to find the volume of each model. If you have blocks or cubes, build each model to check.

a

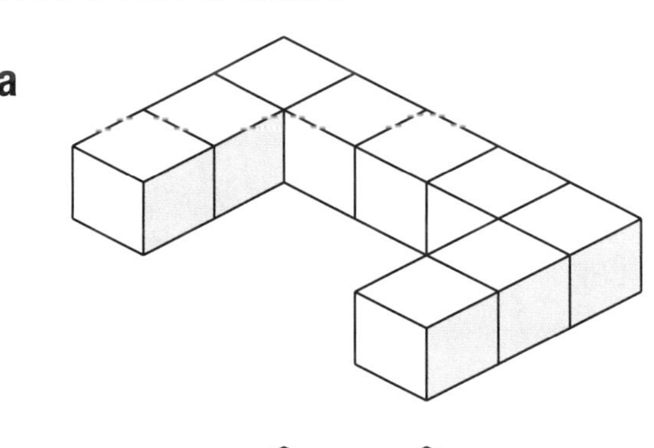

b

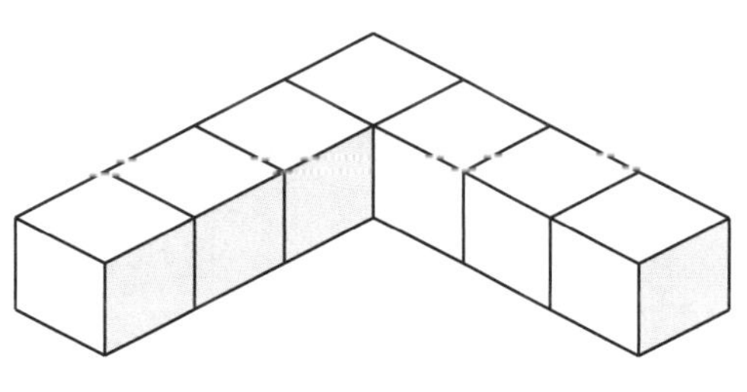

c

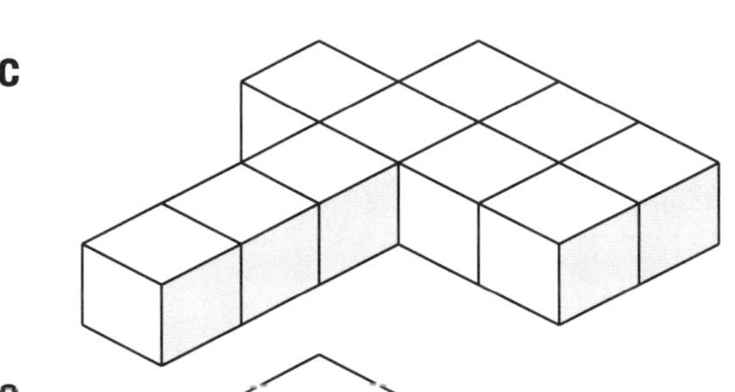

d

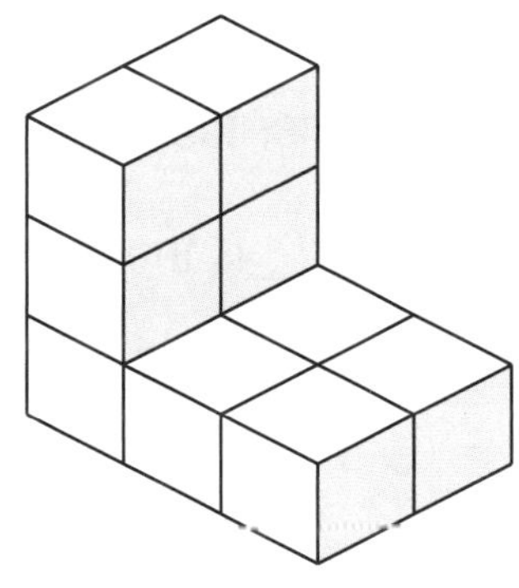

e

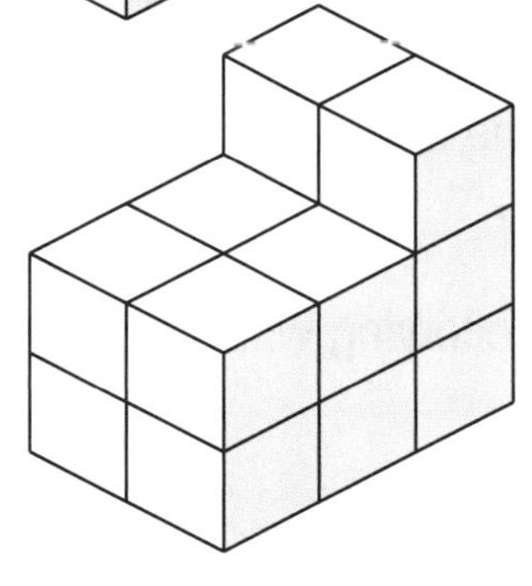

f

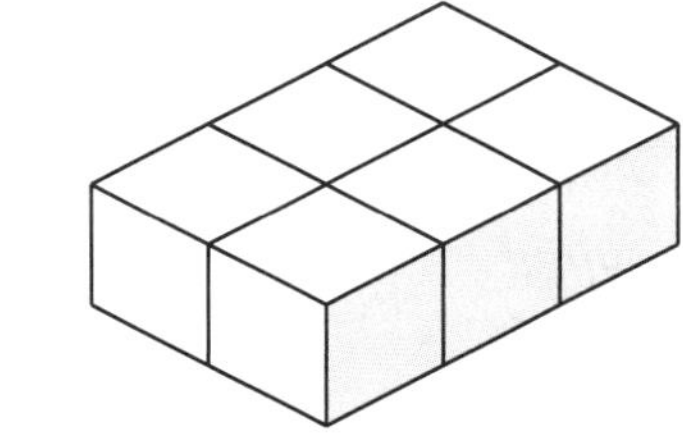

2 Which model uses the most cubes?

3 Which model uses the least cubes?

4 Which two models take up the same space?

Estimate weight of objects using informal units

Use comparison words

1 Copy and complete the following sentences by choosing suitable items from the pictures shown.

a The pineapple is heavier than the __________.

b The pig is lighter than the __________.

c The __________ is heavier than the chicken.

d The __________ is lighter than the car.

e The shell is lighter than the __________.

f The car is heavier than the __________.

g The __________ is the heaviest item.

h The __________ is the lightest item.

i The __________ and the __________ are about the same weight.

j The ______________ is the lightest of the shell, the coconut and the pig.

2 Copy and complete these sentences with one of the following words: *lighter*, *heavier*, *lightest*, *heaviest*

a The baby is __________ than the chicken.

b The chicken is the __________ out of the pig, the car and the chicken.

c The plane is the __________ of all the items.

d The coconut is __________ than the shell.

Recognise formal and informal units of time

Interpret different ways to say time

Copy each clock face and then record all the matching times for each one.

A

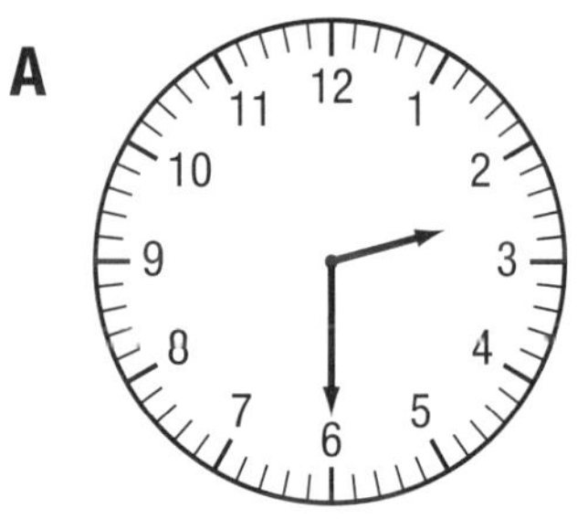

D

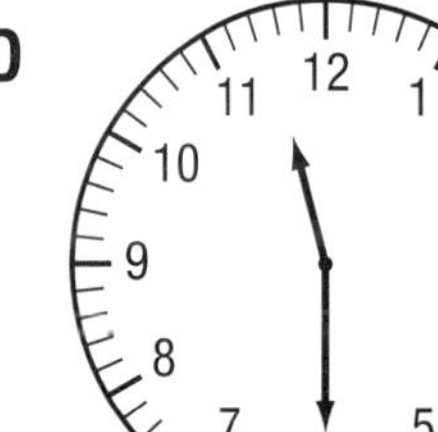

ten o'clock

11:30

six-thirty

10:00

4:00

6:30

half-past 2

half-past 6

seven o'clock

half-past 11

2:30

7:00

two-thirty

four o'clock

eleven-thirty

B

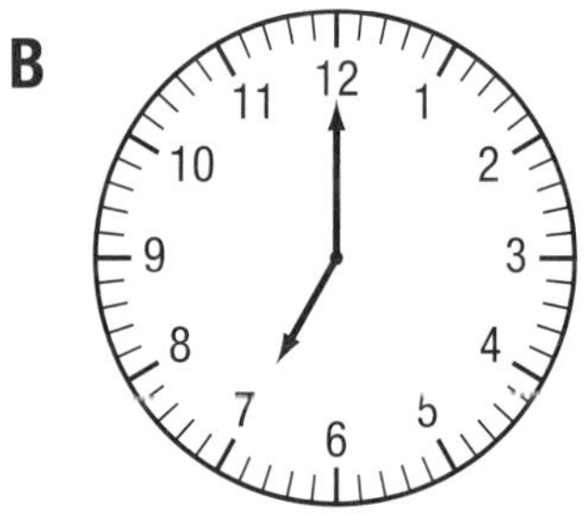

E

C

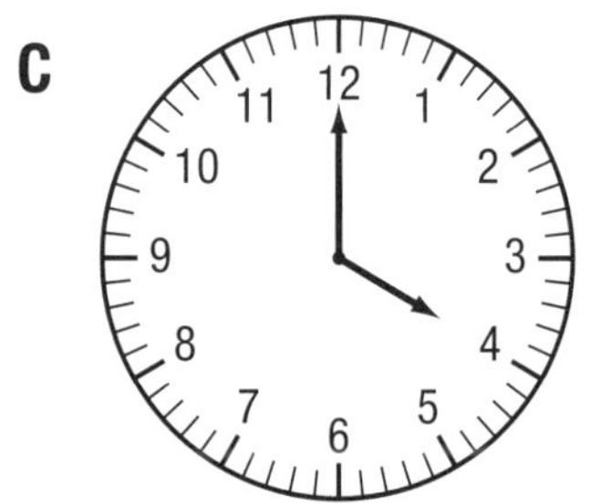

F

Match analogue and digital times

Write the digital time that matches each clock face.

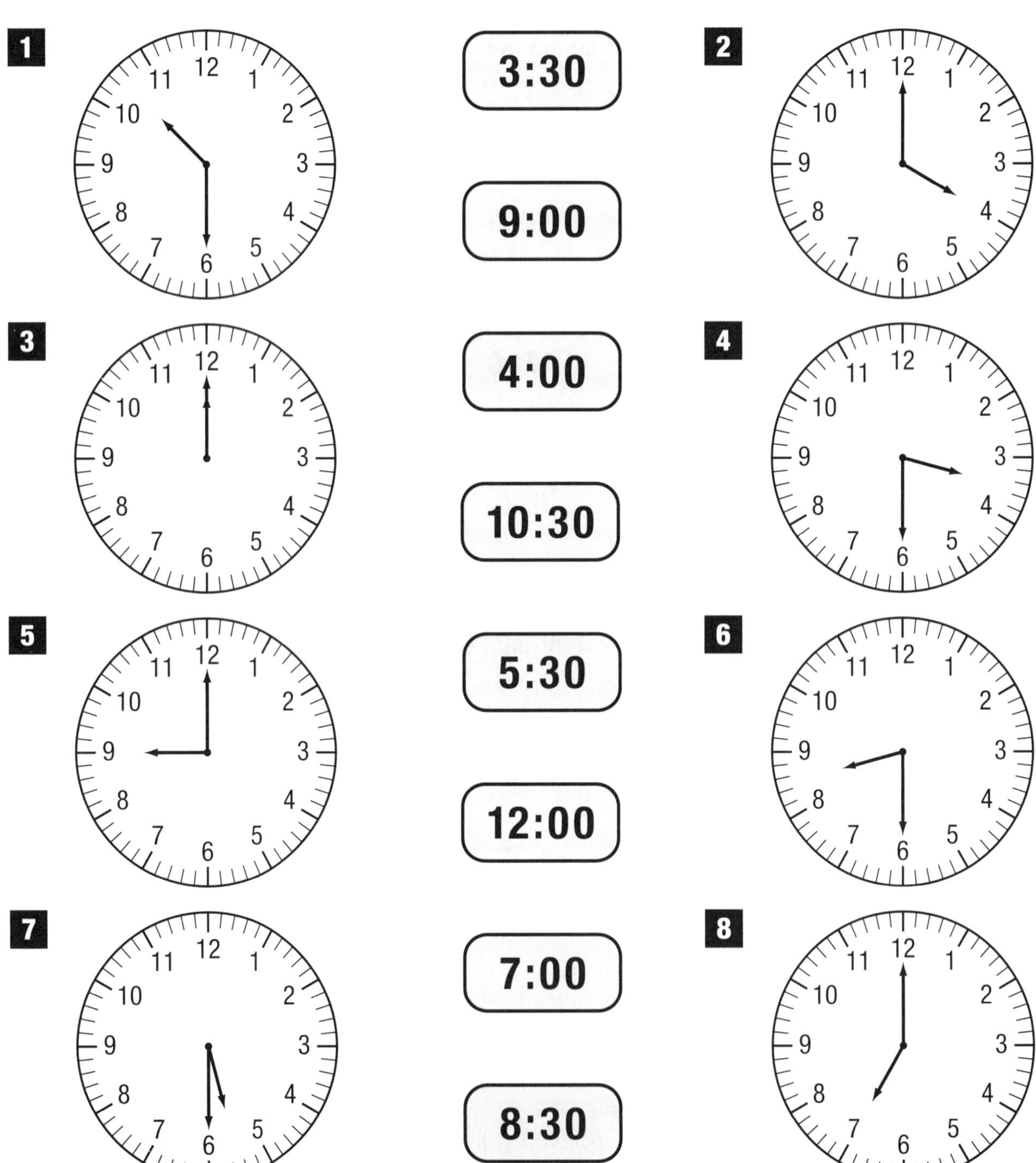

Read and write o'clock times

Write the time from each clock face in analogue and digital form. For example, the first clock shows 7 o'clock or 7:00.

Read and write half-past times

Write the time from each clock face in analogue and digital form. For example, the first clock shows half-past 5 or 5:30.

Order units of time

1 Which of these periods of time is longer?

a 5 days or 1 week

b 15 days or 2 weeks

c 30 minutes or 1 hour

d 1 hour or 75 minutes

e 18 months or 1 year

f 2 years or 20 months

g 1 week or 9 days

h 2 hours or 100 minutes

i 1 year or 10 months

j 2 weeks or 19 days

k 37 months or 3 years

l 150 minutes or 2 hours

m 18 days or 3 weeks

n 5 weeks or 50 days

o 3 hours or 160 minutes

p 10 years or 110 months

q 1 month or 6 weeks

r 40 days or 1 month

s 200 days or 1 year

t 1 year or 70 weeks

u 5 weeks or 2 months

v 30 hours or 1 day

2 Put these periods of time in order from shortest to longest.

a	2 weeks	10 days	1 month	200 hours
b	1 day	20 hours	1000 minutes	12 hours
c	1 year	16 months	40 weeks	2 years
d	30 months	3 years	2 years	100 days
e	300 days	6 months	1 year	12 weeks

Read and interpret a calendar

This calendar is for the year 2015.

January

Su	M	Tu	W	Th	F	Sa
				1	2	3
4	5	6	7	8	9	10
11	12	13	14	15	16	17
18	19	20	21	22	23	24

February

Su	M	Tu	W	Th	F	Sa
1	2	3	4	5	6	7
8	9	10	11	12	13	14
15	16	17	18	19	20	21
22	23	24	25	26	27	28

March

Su	M	Tu	W	Th	F	Sa
1	2	3	4	5	6	7
8	9	10	11	12	13	14
15	16	17	18	19	20	21
22	23	24	25	26	27	28

April

Su	M	Tu	W	Th	F	Sa
			1	2	3	4
5	6	7	8	9	10	11
12	13	14	15	16	17	18
19	20	21	22	23	24	25

May

Su	M	Tu	W	Th	F	Sa
					1	2
3	4	5	6	7	8	9
10	11	12	13	14	15	16
17	18	19	20	21	22	23
24	25	26	27	28	29	30

June

Su	M	Tu	W	Th	F	Sa
	1	2	3	4	5	6
7	8	9	10	11	12	13
14	15	16	17	18	19	20
21	22	23	24	25	26	27
28	29	30				

July

Su	M	Tu	W	Th	F	Sa
			1	2	3	4
5	6	7	8	9	10	11
12	13	14	15	16	17	18
19	20	21	22	23	24	25
26	27	28	29	30	31	

August

Su	M	Tu	W	Th	F	Sa
						1
2	3	4	5	6	7	8
9	10	11	12	13	14	15
16	17	18	19	20	21	22
23	24	25	26	27	28	29

September

Su	M	Tu	W	Th	F	Sa
		1	2	3	4	5
6	7	8	9	10	11	12
13	14	15	16	17	18	19
20	21	22	23	24	25	26

October

Su	M	Tu	W	Th	F	Sa
				1	2	3
4	5	6	7	8	9	10
11	12	13	14	15	16	17
18	19	20	21	22	23	24

November

Su	M	Tu	W	Th	F	Sa
1	2	3	4	5	6	7
8	9	10	11	12	13	14
15	16	17	18	19	20	21
22	23	24	25	26	27	28

December

Su	M	Tu	W	Th	F	Sa
		1	2	3	4	5
6	7	8	9	10	11	12
13	14	15	16	17	18	19
20	21	22	23	24	25	26

Answer these questions about the calendar.

1 How many Tuesdays are in May?

2 How many Sundays are in August?

3 How many weekends are in June?

4 Which months have 31 days?

5 Which months have 30 days?

6 How many days are in February?

7 On what day does April end?

8 On what day does July start?

9 What day is the 17th August?

10 What day is the 20th October?

11 What day is the 13th March?

12 What day is the 15th May?

13 What date is the third Friday in September?

14 What date is the second Sunday in November?

15 What date is the fourth Saturday in May?

16 What is the total number of days in the first 4 months of the year?

17 What is the total number of days in the last 4 months of the year?

Assessment **Measurement**

1 Measure and record the length of each line in centimetres.

a

b

c

d

2 Measure and record the perimeter of each shape.

a

b
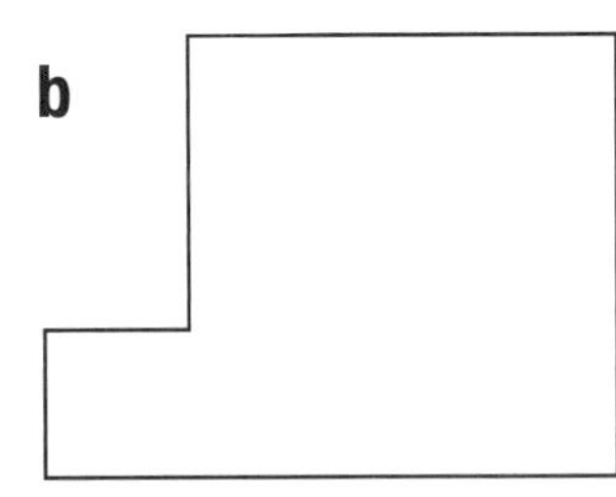

c
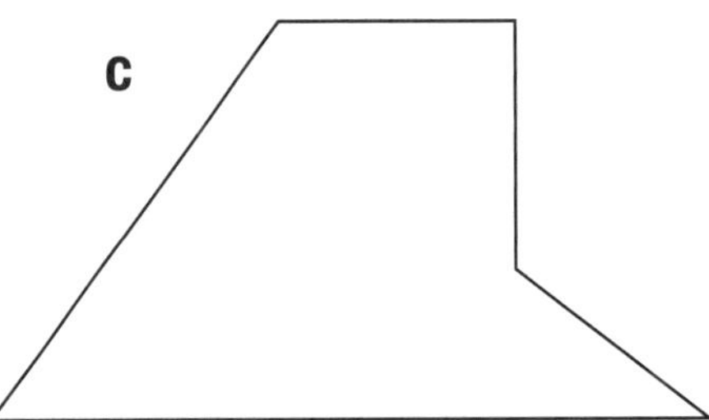

3 Answer the questions below the shapes.

A
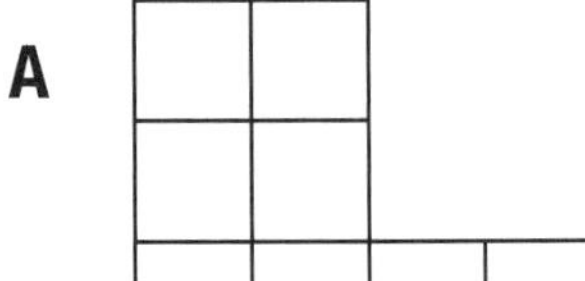

B
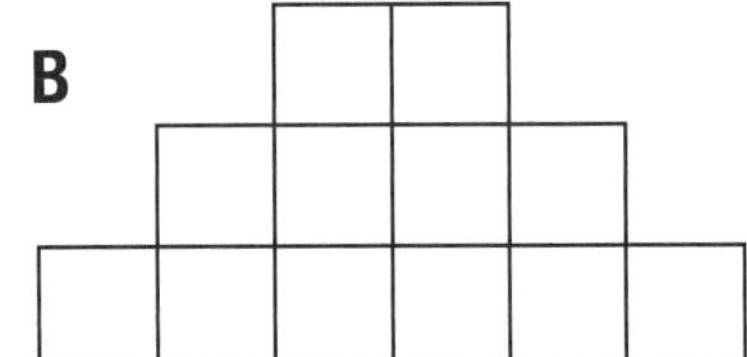

C
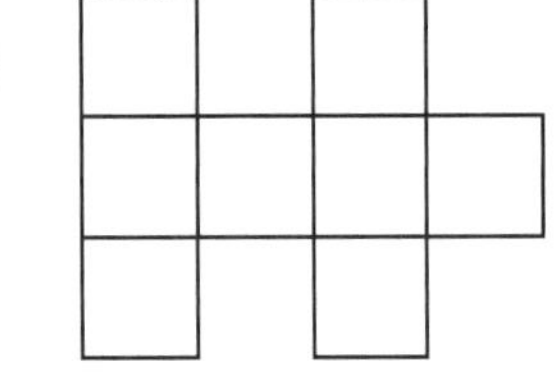

a Which shape has the largest area?

b How many square units larger than C is B?

c If A and B were joined, what area would they cover?

4 Answer the questions below the models.

A

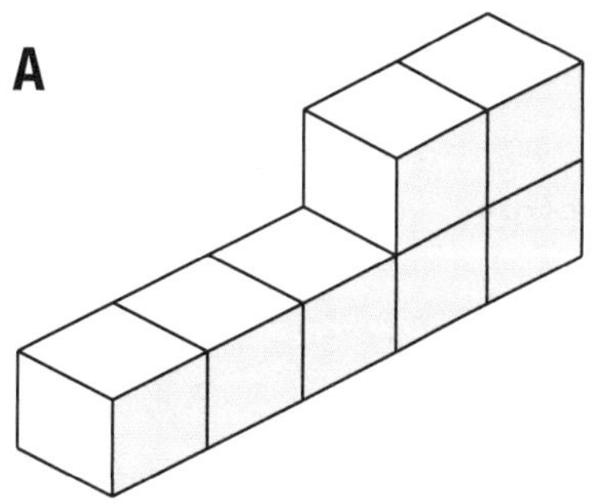

B

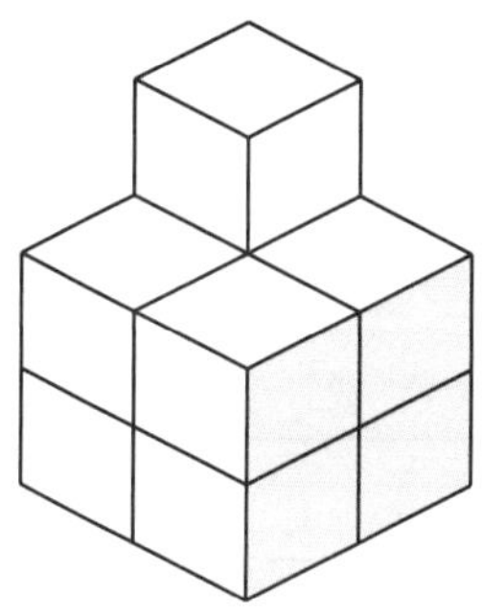

C

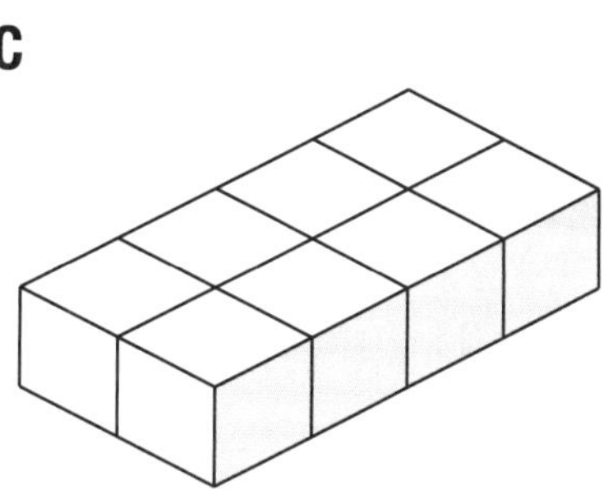

a Which model has the largest volume?

b How many cubes smaller than B is A?

c If B and C were joined, what would their volume be?

5 Copy and complete this chart.

	Words	Analogue time	Digital time
a	six o'clock	6 o'clock	
b		$\frac{1}{2}$ past 9	
c			2:30
d	half-past seven		
e			11:00

6 Write the time from each clock in analogue and digital form.

a 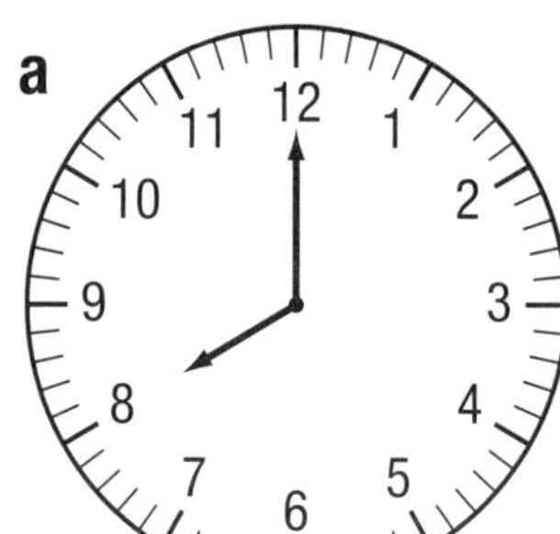

b 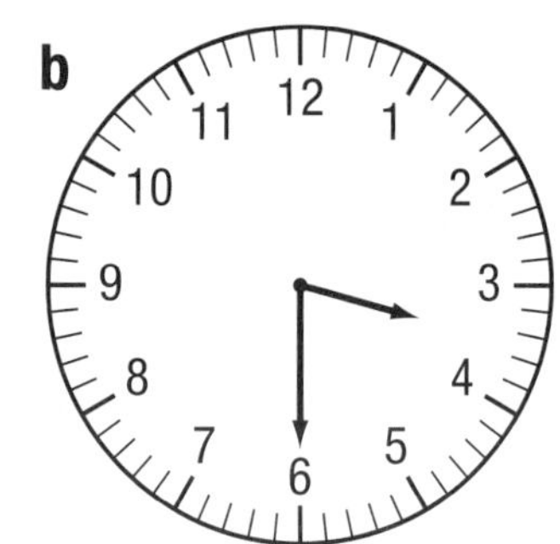

c 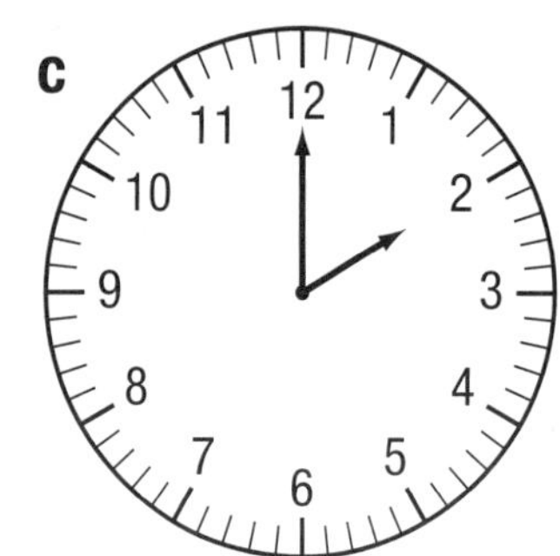

7 Draw clock faces to show these times.

a 5:00 **b** 1:30 **c** 10:00 **d** 12:30

8 Which period of time is longer?

a 1 week or 9 days

b 50 minutes or 1 hour

c 2 years or 20 months

d 6 weeks or 1 month

e 11 days or 2 weeks

f 22 hours or 1 day

g 450 days or 1 year

h 15 months or 1 year

i 2 hours or 90 minutes

j 2 years or 150 weeks

k 40 days or 2 months

l 3 days or 100 hours

Strand Space and Shape

Identify two and three-dimensional shapes

Match two-dimensional shapes with their names

Write the name of each shape, and then write the letter of the matching shapes.

square

circle

triangle

rectangle

hexagon

trapezium

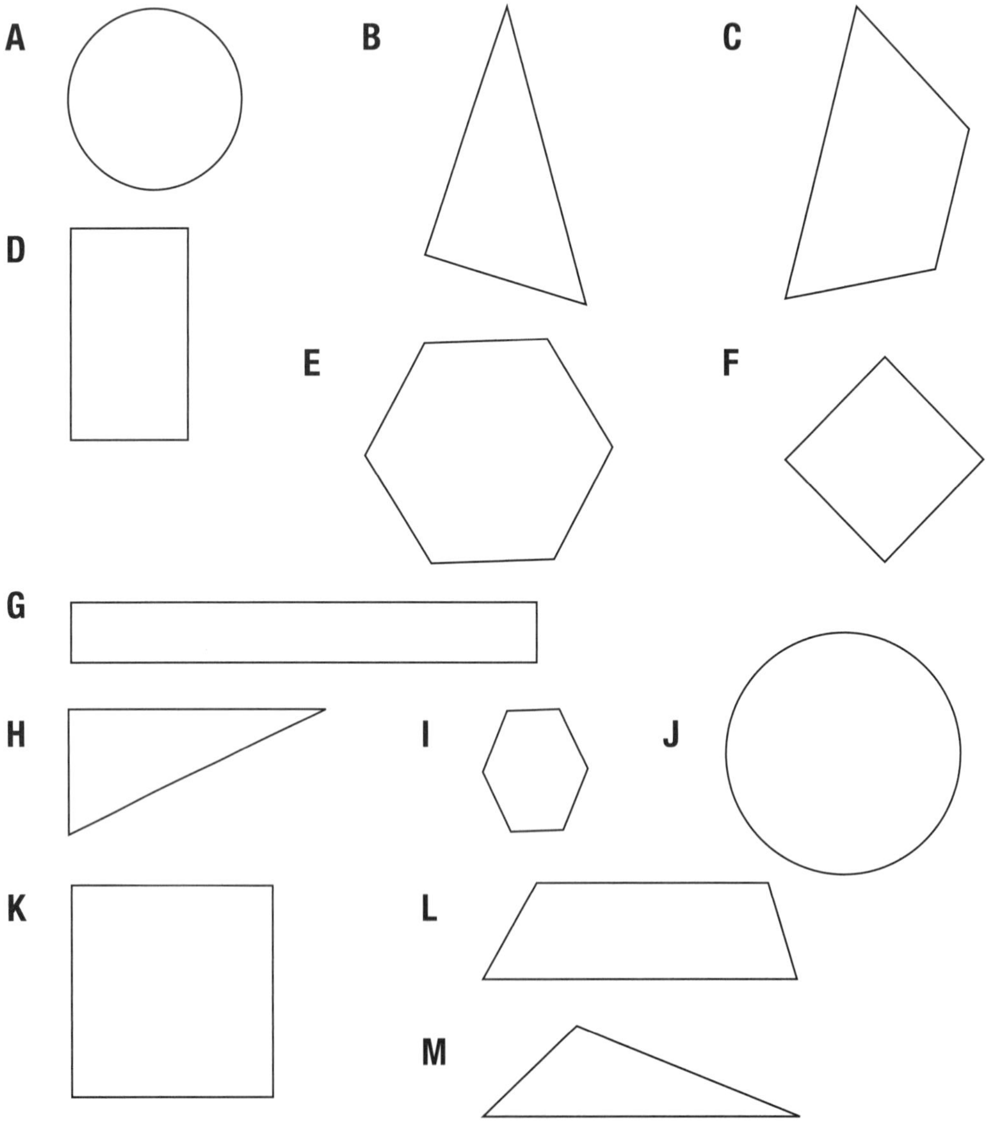

A

B

C

D

E

F

G

H

I

J

K

L

M

Match three-dimensional shapes with their names

Write the name of each shape, and then write the letter of the matching shapes.

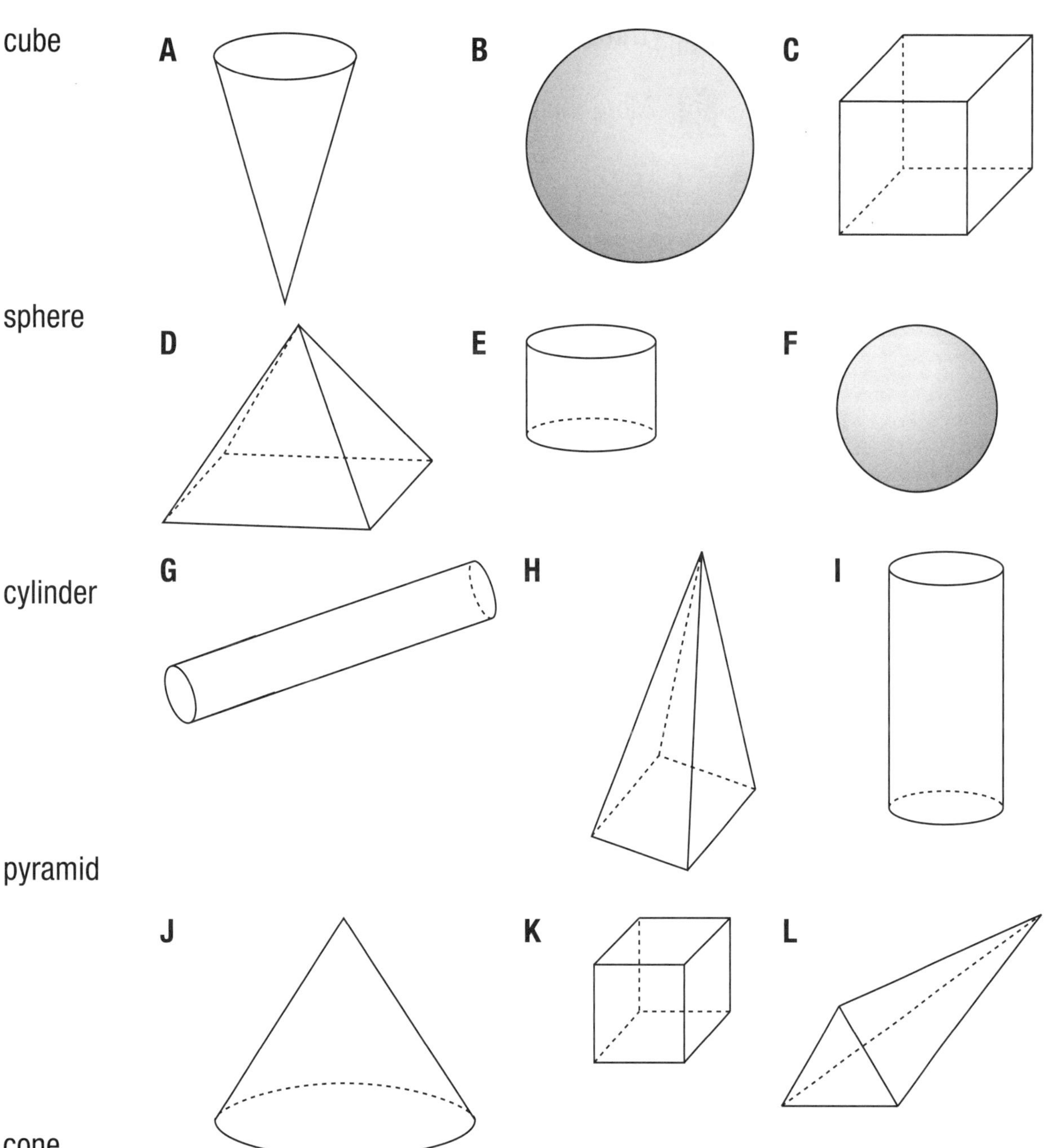

Identify features of two-dimensional shapes

Which of the shapes below have:

1 Four corners?

2 Three straight sides?

3 Four equal sides?

4 No straight sides?

5 More than four corners?

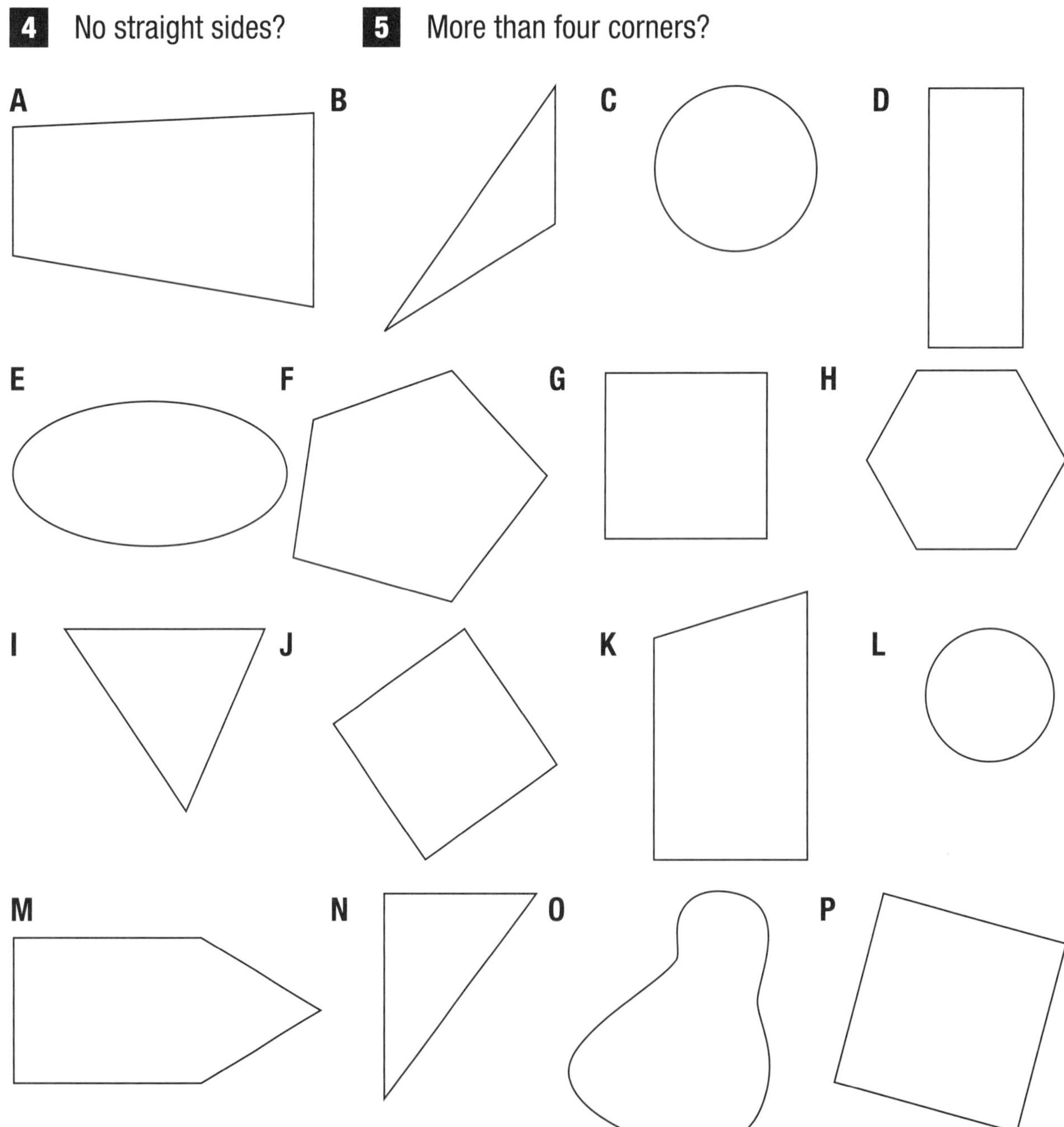

Identify features of three-dimensional shapes

Which of the shapes below have:

1 Eight corners?

2 Six faces?

3 Eight edges?

4 A curved surface?

5 Five faces?

6 Five corners?

A

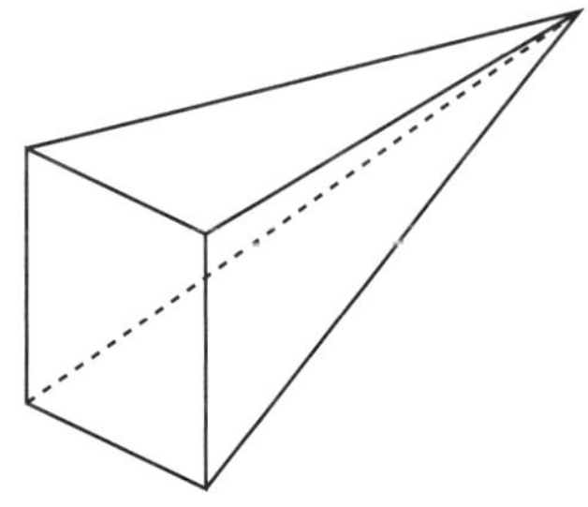

B

C

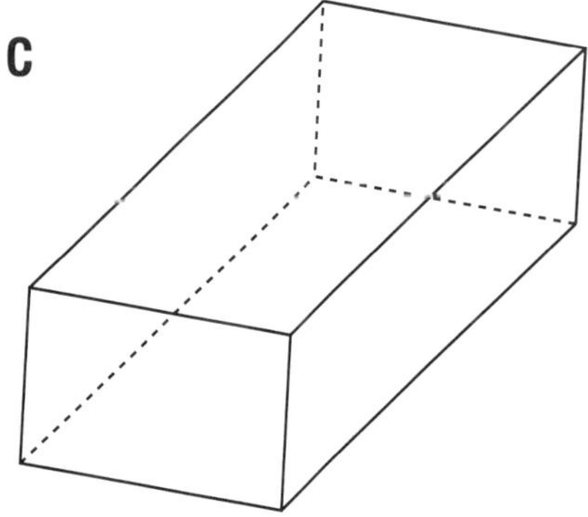

D

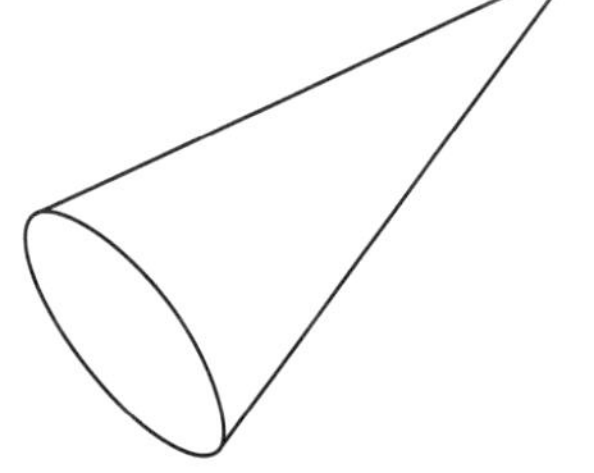

E

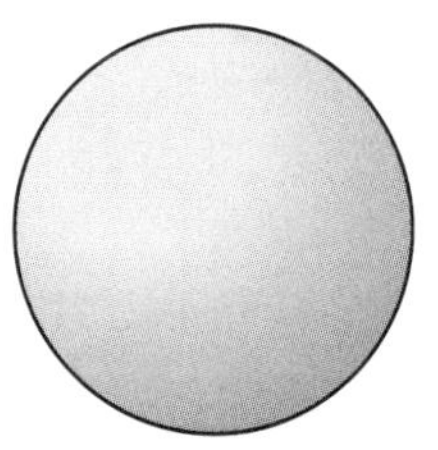

F

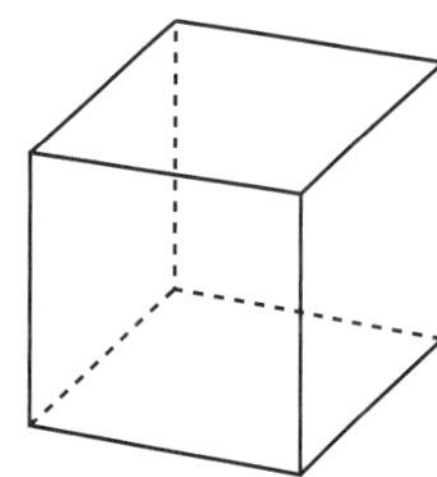

G

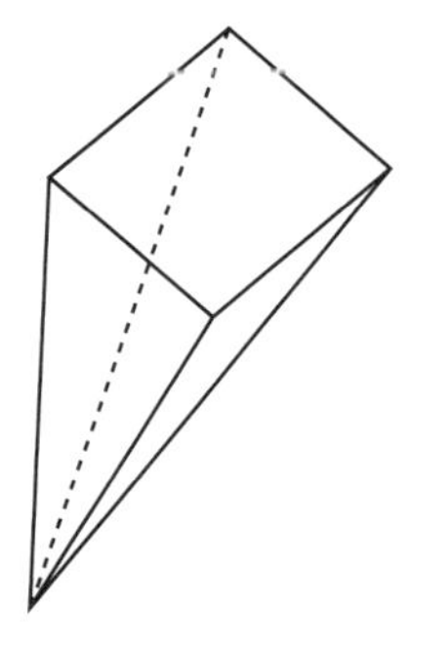

H

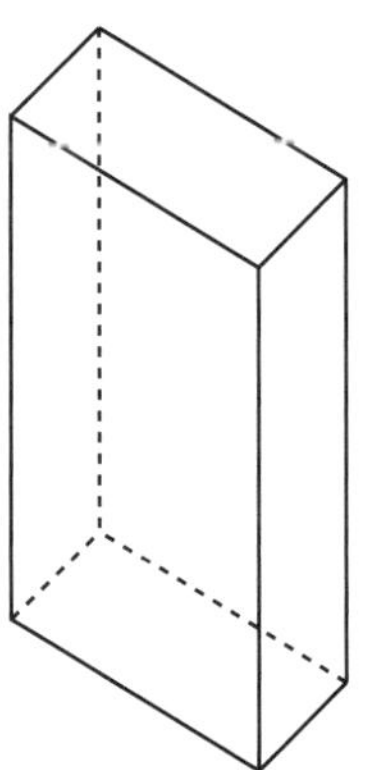

I

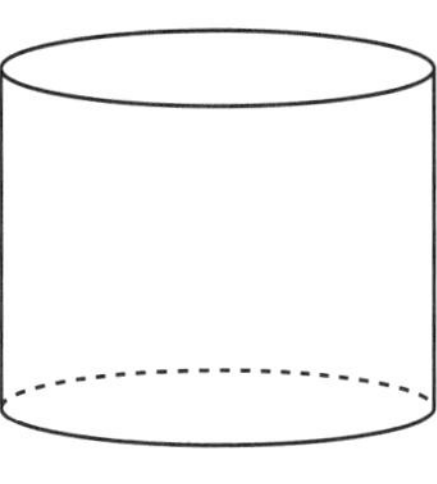

J

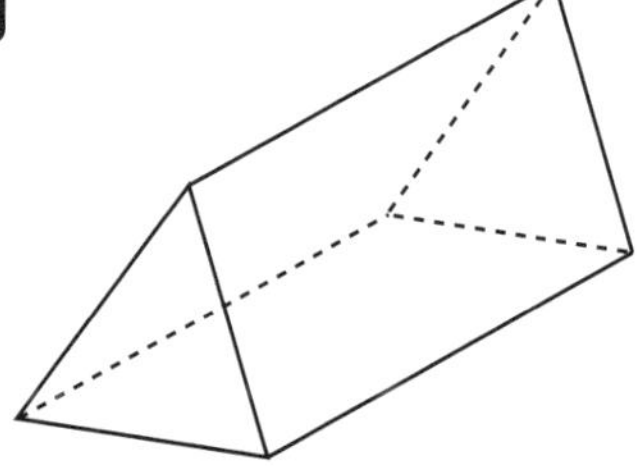

K

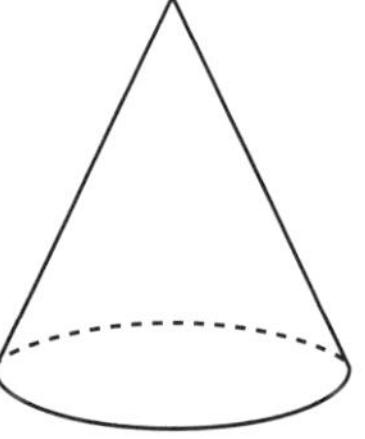

L

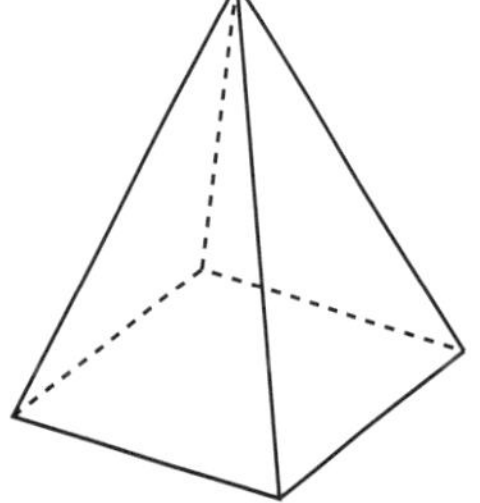

Recognise, compare and order angles

Make $\frac{1}{4}$, $\frac{1}{2}$, $\frac{3}{4}$ and full turns

If the long hand on each clock face makes the turn shown above the clock face, what number will it be pointing at?

1 half turn

2 quarter turn

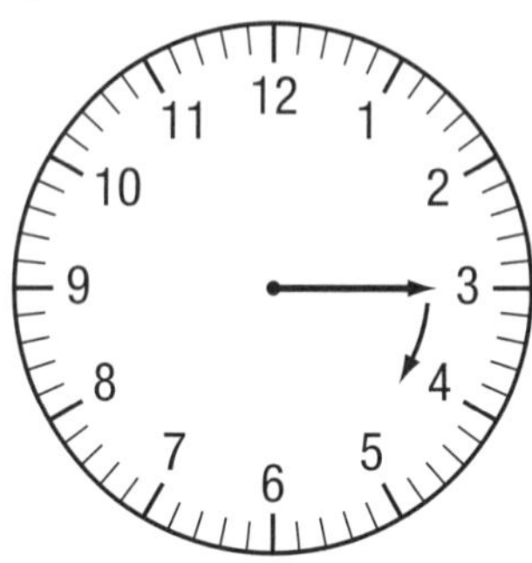

3 half turn

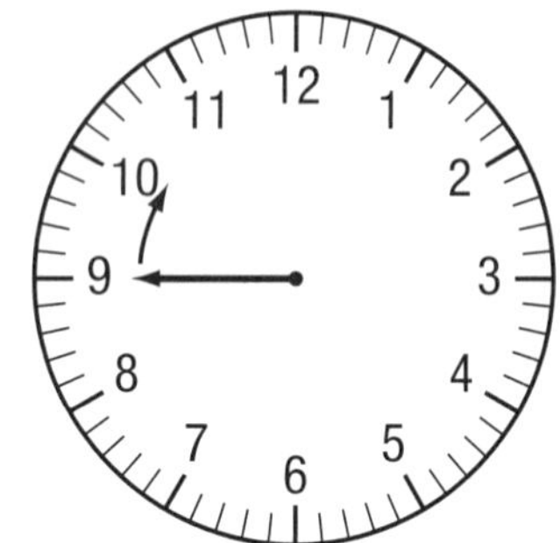

4 three-quarter turn

5 full turn

6 quarter turn

7 full turn

8 three-quarter turn

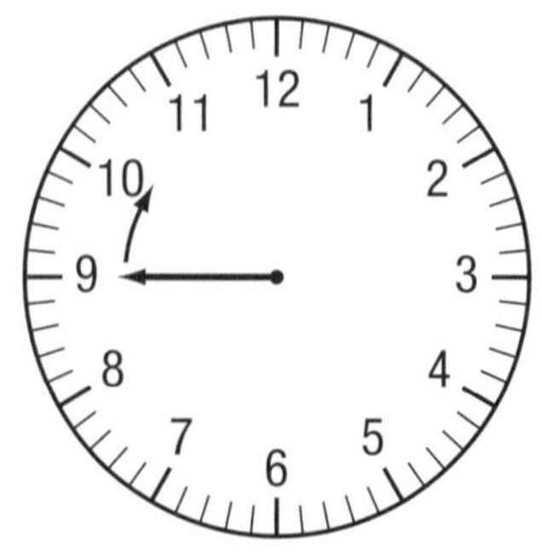

9 half turn

Name and use the four main directions

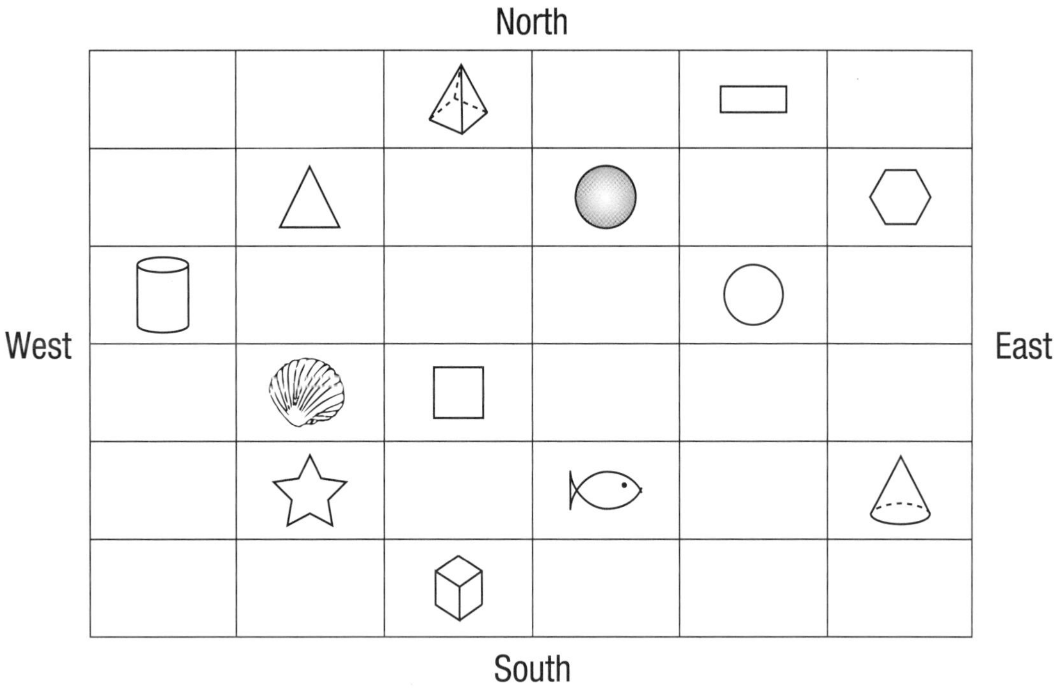

North

West East

South

Copy the sentences and complete the missing directions.

1 The triangle is ______ of the star.

2 The circle is ______ of the rectangle.

3 The cone is ______ of the star.

4 The cylinder is ______ of the circle.

5 The pyramid is ______ of the rectangle.

6 The cube is ______ of the square.

7 The hexagon is ______ of the cone.

8 The fish is ______ of the sphere.

9 The square is ______ of the pyramid and ______ of the cube.

10 The sphere is ______ of the hexagon and ______ of the triangle.

11 The fish is ______ of the star and ______ of the sphere.

12 The shell is ______ of the star and ______ of the square.

Identify right and straight angles

1 Which of these angles are right (quarter turn) angles?

a 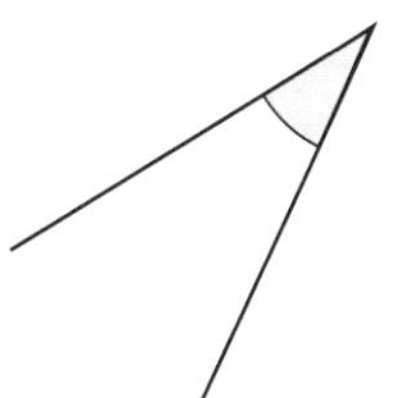 **b** 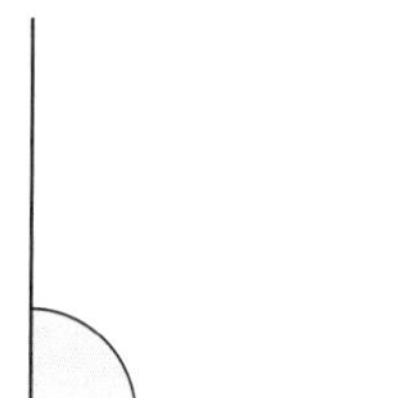 **c** 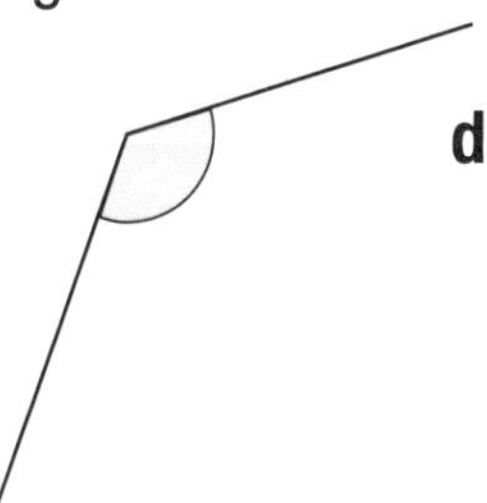 **d** 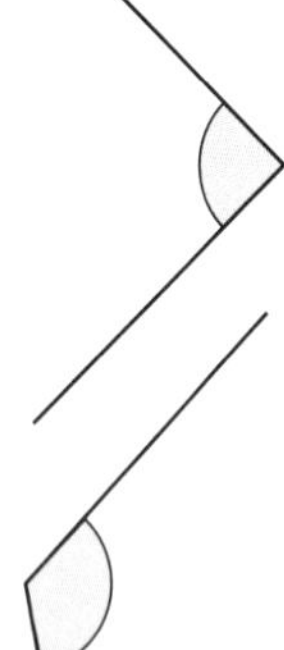

e 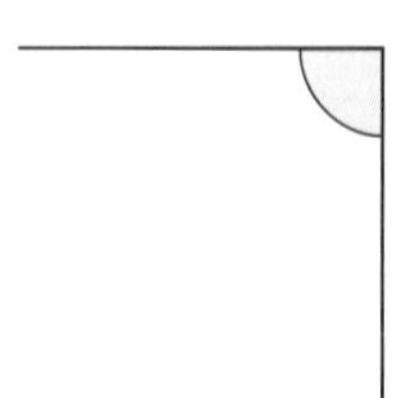 **f**  **g** 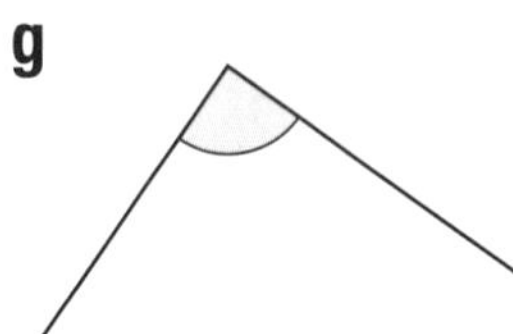 **h** 

2 Which of these angles are straight (half turn) angles?

a 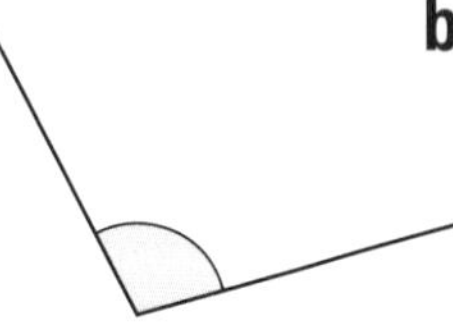 **b** 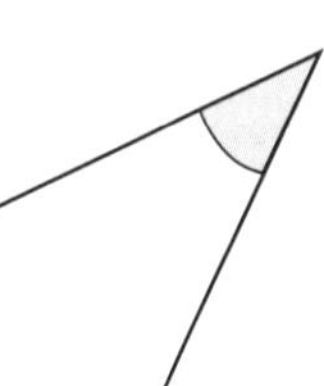 **c** 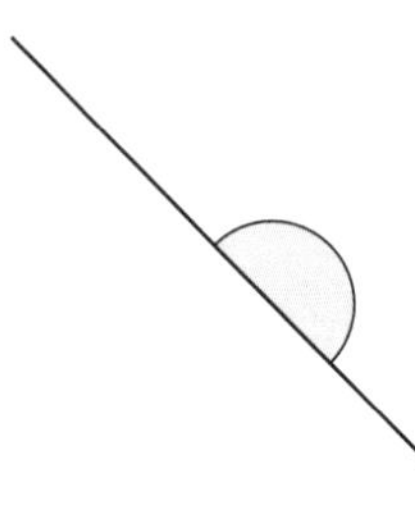 **d** 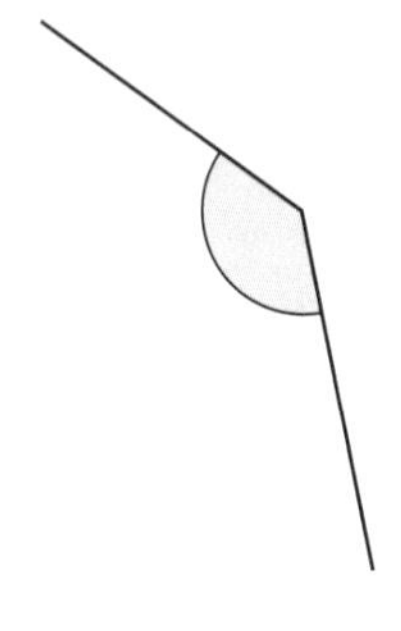

e 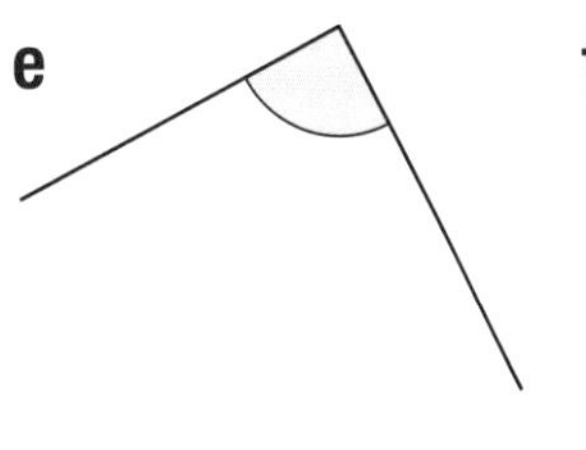 **f** 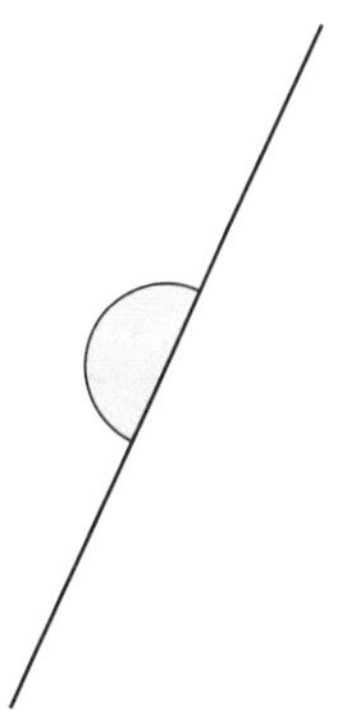 **g** 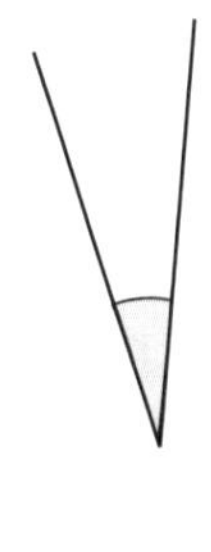 **h**

Assessment — Space and Shape

1 Which of these shapes is *not* a circle?

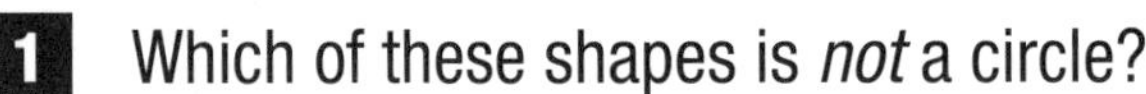

a 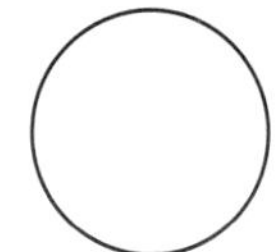

b 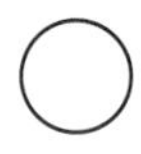

c 

d 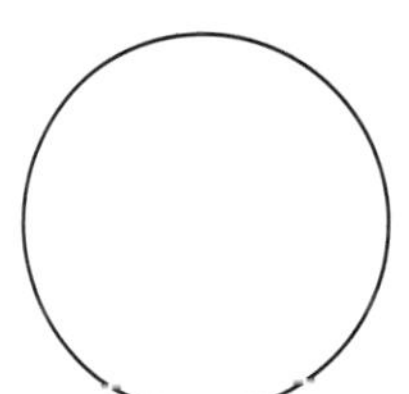

2 Which of these shapes is *not* a square?

a 

b 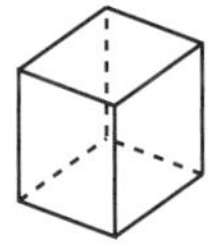

c 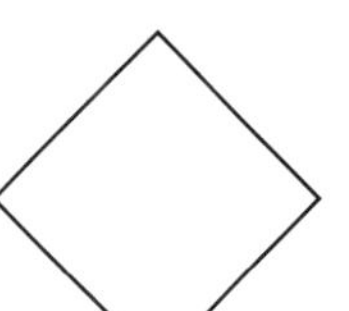

d 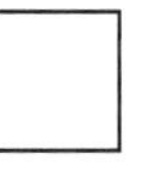

3 Which of these shapes is *not* a cylinder?

a 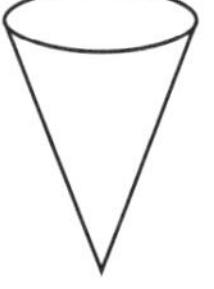

b 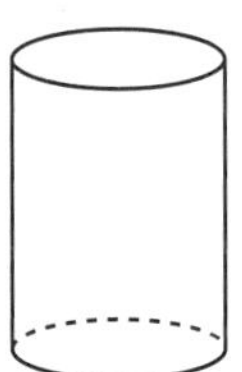

c 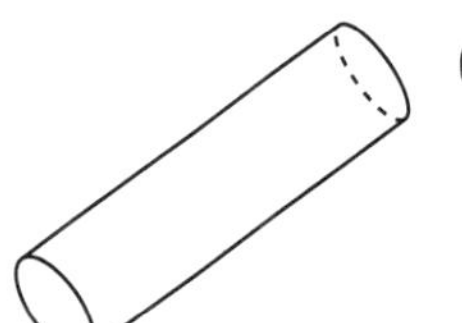

d 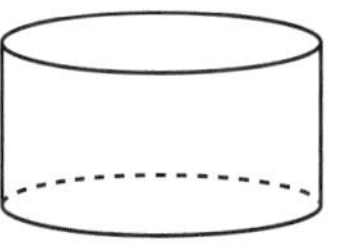

4 Which of these shapes is *not* a pyramid?

a 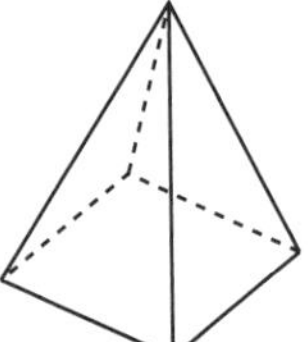

b 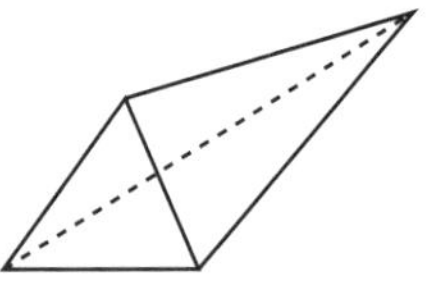

c 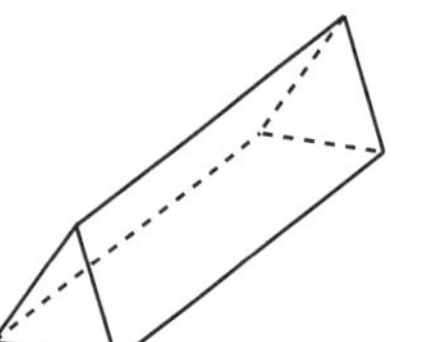

d 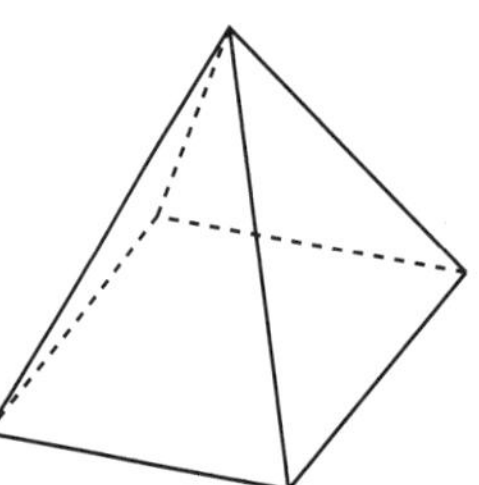

5 Which of these shapes have four corners?

a 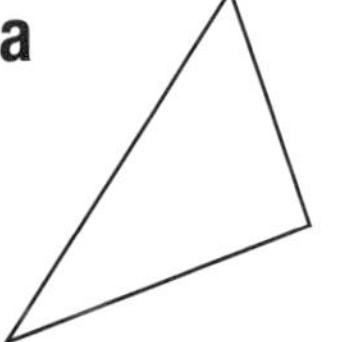

b 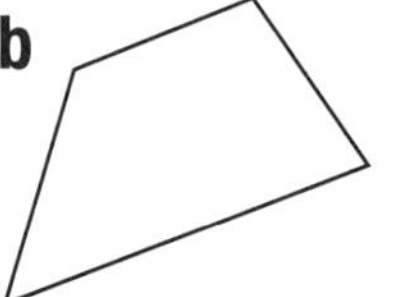

c 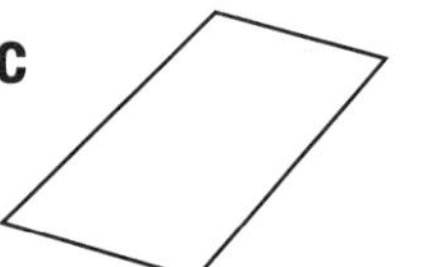

d 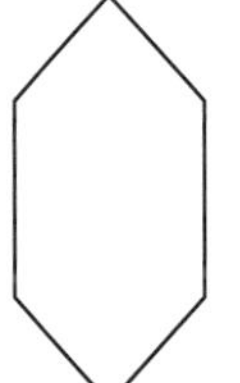

e 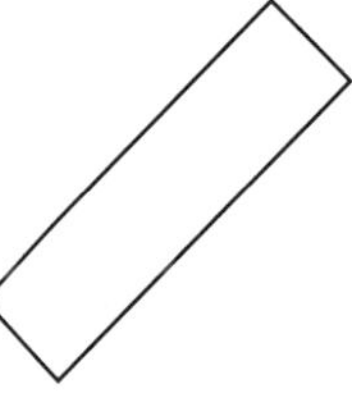

6 Which of these shapes have six faces?

a 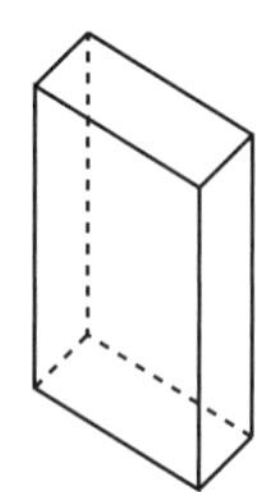

b 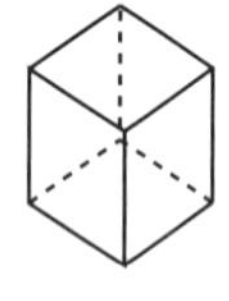

c 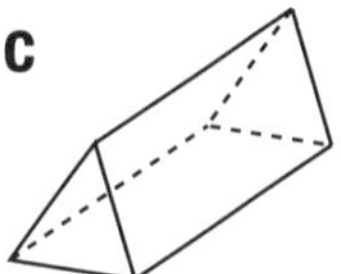

d 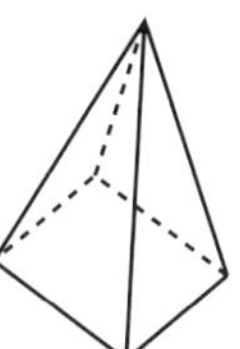

e 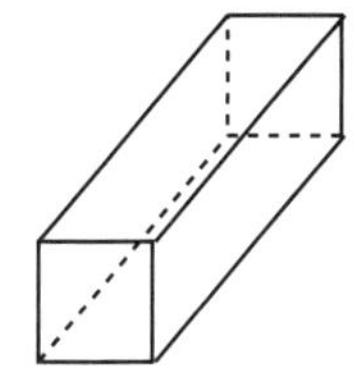

7 If the hand on each clock face makes the turn shown above the clock face, what number will it be pointing at?

a three-quarter turn

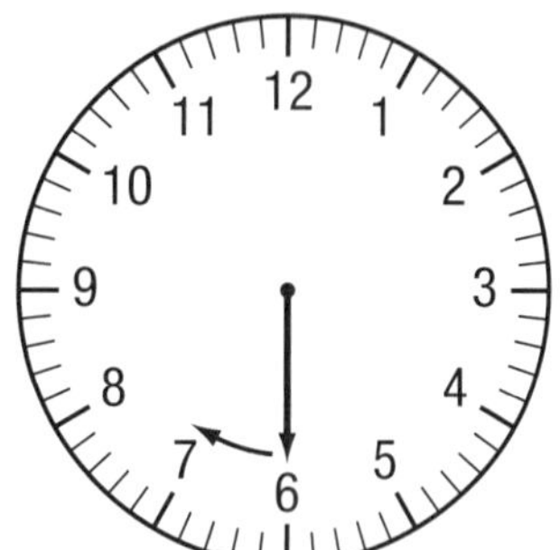

b half turn

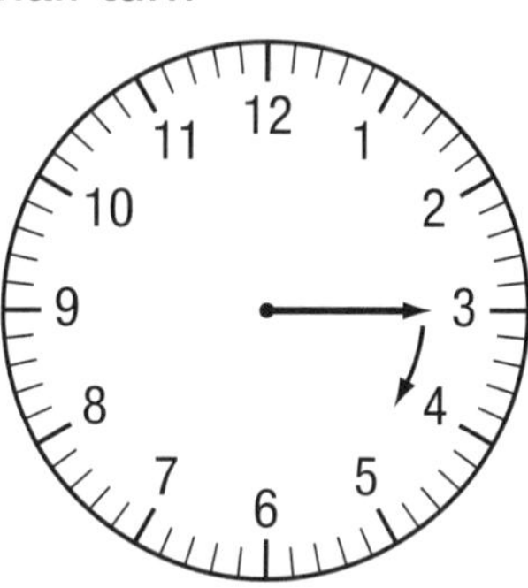

c quarter turn

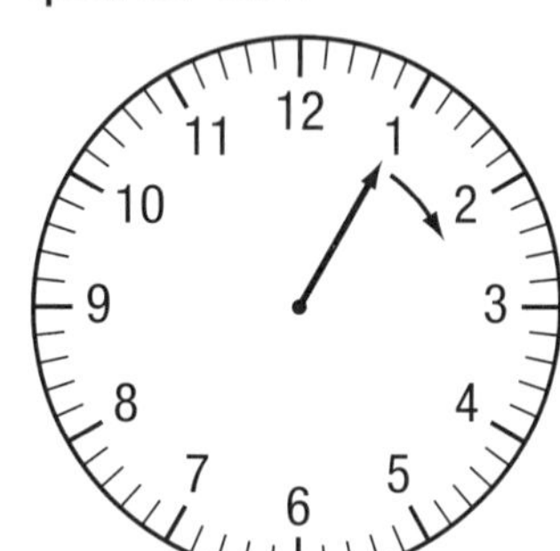

8 Complete the missing direction in each sentence.

a G is ______ of Z.

b D is ______ of P.

c B is ______ of Z.

d A is ______ of T.

e P is ______ of S.

9 Which of these angles are right angles?

a **b**

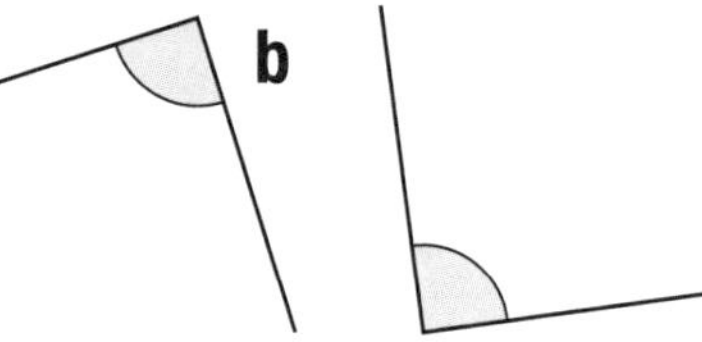

c

d

e

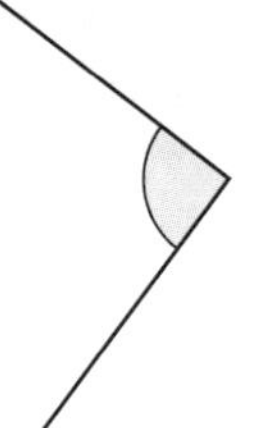

10 Which of these angles are straight angles?

a

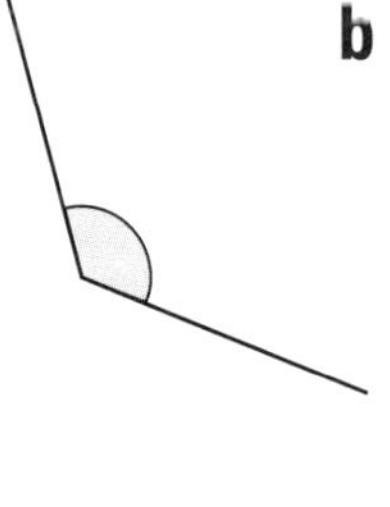

b

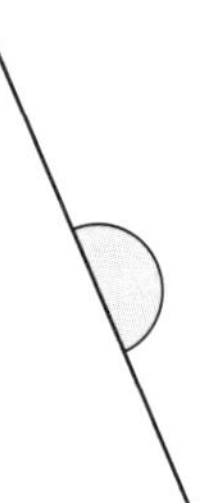

c **d**

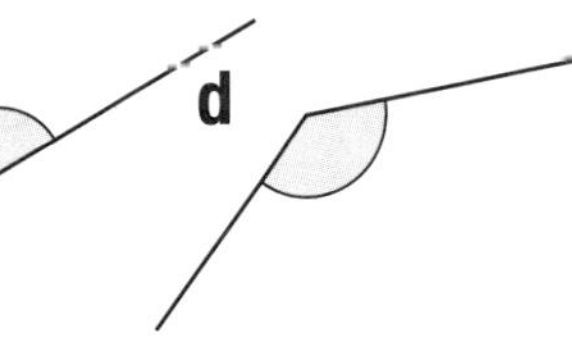

e

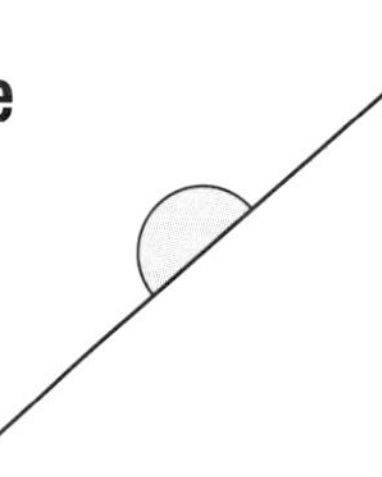

Strand **Chance and Data**

Use the language of chance to describe everyday events

Describe likelihood

Choose one of the following words to describe each situation:

certain **possible** **impossible**

1 It will rain bats and balls.	**2** A pig will write a letter.
3 I will go to sleep.	**4** A baby will drive a PMV.
5 I will walk with books balanced on my head.	**6** The dice will land on 5.

Predict outcomes for an event

You need a coin and some counters.

Toss the coin. If it lands with 'heads' facing up, put a counter on the circle closest to START on the 'heads' side. If it lands with 'tails' facing up, put a counter on the circle closest to START on the 'tails' side.

Toss the coin ten times. How many counters are on each side? Are there more heads, more tails or are there the same number of each?

Repeat this activity five times. Before you start, predict how many times the 'heads' side will win and how many times the 'tails' side will win and how many times they will draw.

Heads

Start

Tails

Draw and interpret simple graphs

Read and interpret information using tallies

1 Copy the table and complete the column titled 'Total'.

a How many people chose number 3?

b Which number was most popular?

c How many people were surveyed?

d Which number was least popular?

e How many more people chose number 5 than number 9?

f How many fewer people chose number 8 than number 7?

g Which number was chosen by 11 people?

Favourite number

Number	Tally	Total
1	||	2
2	𝍸 |||	
3	𝍸 𝍸 |	
4	|||	
5	𝍸 𝍸 𝍸	
6	|||	
7	𝍸 𝍸 𝍸 ||||	
8	||||	
9	𝍸 |	

2 Copy the table and complete the column titled 'Total'.

a How many carrots were eaten?

b How many bananas were eaten?

c Which fruit or vegetable was eaten the most?

d How many yams and eggplants were eaten in total?

e How many more bananas were eaten than yams?

Fruit and vegetables eaten by a family in one week

Fruit/Vegetable	Tally	Total
Bananas	𝍸 𝍸 ||	
Pineapples	||	
Carrots	𝍸 𝍸	
Yams	𝍸 ||||	
Melons	||	
Eggplants	||||	
Coconuts	|||	

Read picture graphs

1 Answer the following questions about the graph.

Eggs collected during one week

Monday	◯◯◯◯◯
Tuesday	◯◯◯◯◯◯
Wednesday	◯◯◯◯
Thursday	◯◯◯◯◯
Friday	◯◯◯◯◯◯◯◯
Saturday	◯◯◯
Sunday	◯◯◯◯◯◯

◯ = 1 egg

a How many eggs were collected on Sunday?

b How many eggs were collected on Monday?

c On which day were the most eggs collected?

d On which day were four eggs collected?

e What was the total number of eggs collected on Wednesday and Thursday?

f How many more eggs were collected on Friday than on Saturday?

g How many more eggs were collected on Sunday than on Wednesday?

h On which days were an even number of eggs collected?

i On which days were the same number of eggs collected?

j How many fewer eggs were collected on Saturday than on Tuesday?

k What was the total number of eggs collected during the week?

l How many eggs were collected on the weekend?

2 Answer the following questions about the graph.

a Which girl threw three goals during the game?

b Who threw the least number of goals?

c How many goals did Kate throw?

d What was the total number of goals thrown by Marie and Grace?

e What was the total number of goals thrown during the game?

f How many more goals did Grace throw than Inala?

g How many more goals did Marie throw than Elsie?

h How many goals did Elsie, Kate and Inala throw together?

i If the opposing team threw 25 goals, how many goals did the above team win by?

j Which two girls threw 13 goals between them?

Read a bar graph

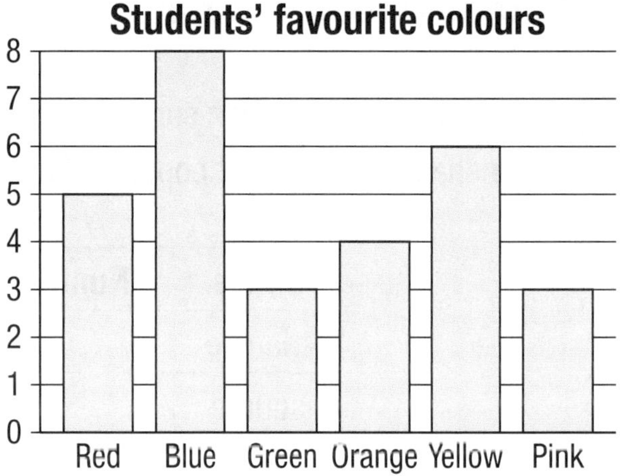

1 How many students chose blue as their favourite colour?

2 How many more students chose red than green?

3 How many more students chose orange than pink?

4 Which two colours were chosen by the same number of students?

5 How many students chose either red or blue?

6 How many more students chose blue than pink?

7 What is the difference in student choices between green and yellow?

8 How many students took part in the survey about favourite colours?

9 Write the colours in order from the most popular to the least popular.

10 If four more students were surveyed and chose red as their favourite colour, what would be the total number of students who chose red?

Collect and record data

1 Copy this table and then survey the students in your grade to find out which of the four shapes is the most popular. Record their choices on the table.

Shape	Number of students	Total
triangle		
square		
circle		
rectangle		

2 Copy this graph and then use the data you have collected above to complete it.

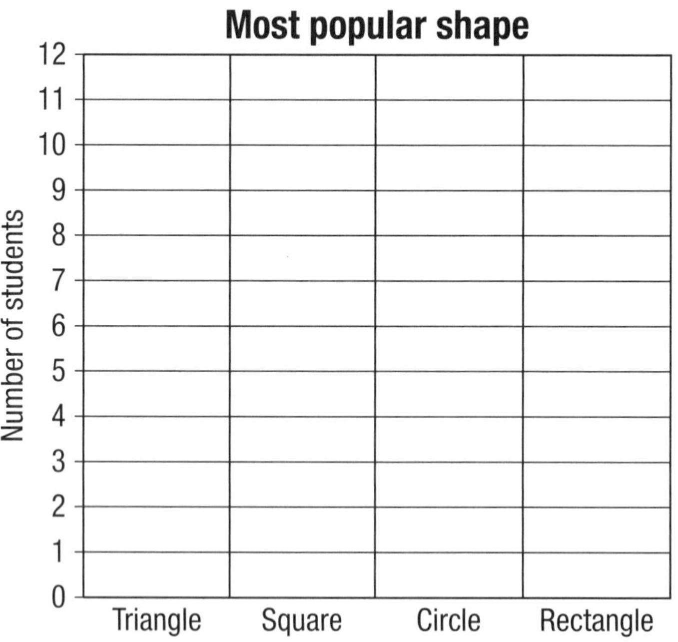

3 Which shape was most popular?

4 How many students took part in the survey?

Assessment Chance and Data

1 Alex counted the coins in his money box and entered the results on this table.

Coins in money box

Type of coin	Tally
K1	𝍸 ||
50t	𝍸 𝍸 ||
20t	𝍸 𝍸 𝍸 |
10t	𝍸 𝍸 |||
5t	𝍸 𝍸 𝍸

a How many K1 coins are in his money box?

b How many 20t coins are in his money box?

c Which coin is there the least of?

d What is the value of the 10t coins?

e How many more 20t coins are there than 10t coins?

2 Mary tossed a dice and graphed how many times it landed on each number.

a How many times did it land on 6?

b How many times did it land on 1?

c How many times did Mary toss the dice?

d Which two numbers did it land on five times?

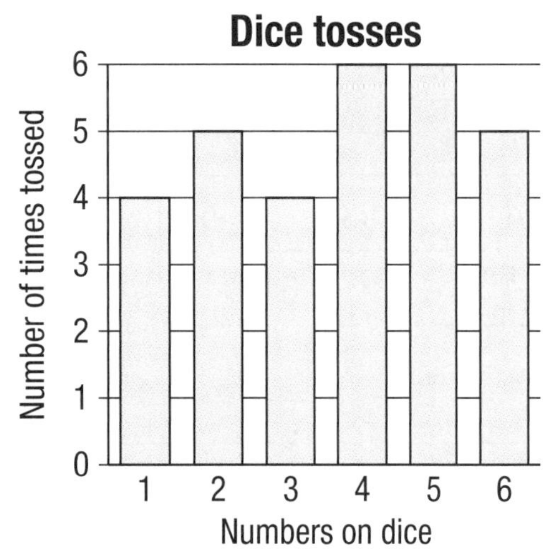

e How many more times did it land on 4 than on 3?

f How many numbers are on the dice?

Strand Patterns

Make and describe patterns and number patterns

Predict missing numbers

Copy and complete the number charts.

1

21	22	23	
31	32	33	
41	42		

2

33					
	44				
53		55			58
	64				68

3

	32			
51		53		55
61			64	

4

	86	87		89
		107		
	116			

5

	272				
281					

6

				529
		537		

Complete pattern grids

Copy and complete the pattern grids.

1

2

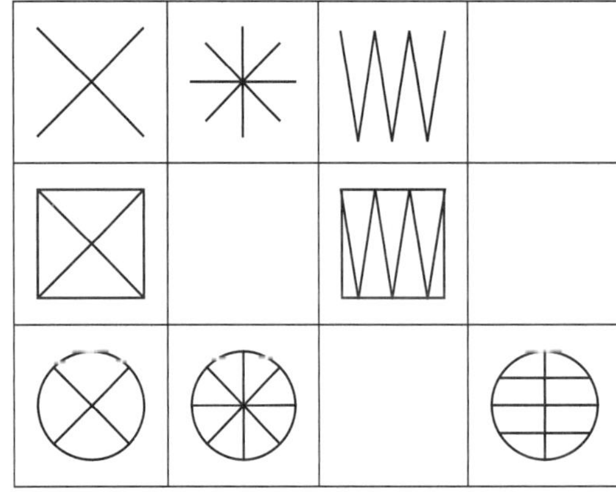

3

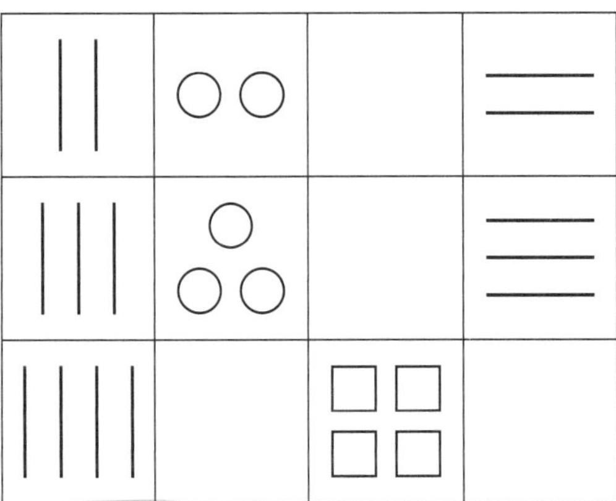

4

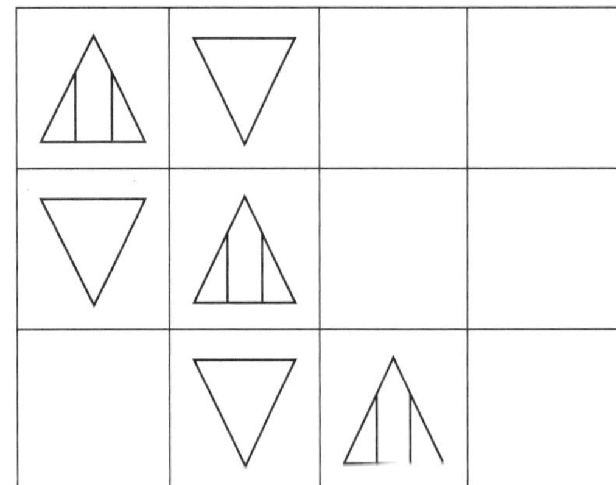

5

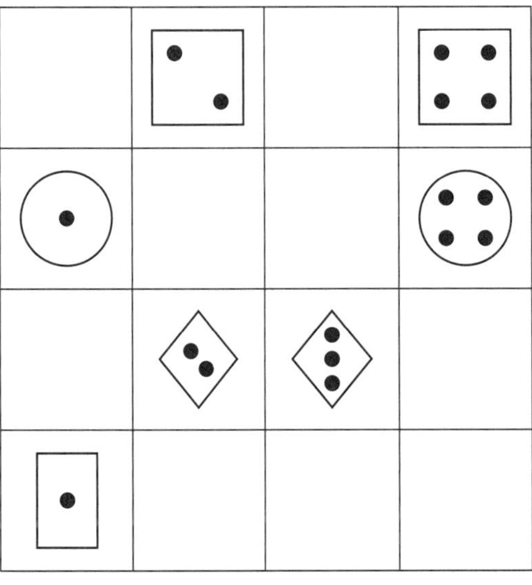

6

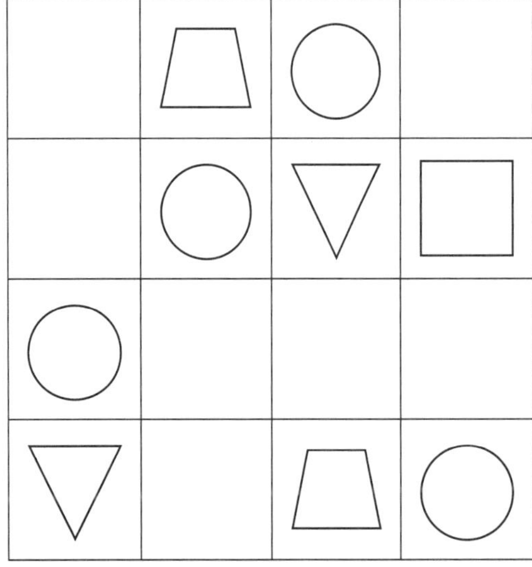

Complete counting sequences

Copy and complete the counting sequences.

1. 35, 34, 33, _____, _____, _____, _____, _____, _____, _____
2. 50, 52, _____, _____, 58, _____, _____, _____, _____, _____
3. 44, 49, 54, _____, _____, _____, _____, _____, _____, _____
4. 19, 21, 23, _____, _____, _____, _____, _____, _____, _____
5. 23, 33, 43, _____, _____, _____, _____, _____, 103, 113, _____
6. 58, 63, 68, _____, _____, _____, _____, _____, _____, _____
7. 67, 65, 63, _____, _____, _____, _____, _____, _____, _____
8. 151, 141, 131, _____, _____, _____, _____, _____, _____, _____
9. 73, _____, 77, 79, _____, _____, _____, _____, _____, _____
10. 92, 87, 82, _____, _____, _____, _____, _____, _____, _____
11. 100, 102, 104, _____, _____, _____, _____, _____, _____
12. 160, 155, 150, _____, _____, _____, _____, _____, _____
13. 200, 210, 220, _____, _____, _____, _____, _____, _____
14. 235, 225, 215, _____, _____, _____, _____, _____, _____
15. 187, 185, 183, _____, _____, _____, _____, _____, _____
16. 284, 289, 294, _____, _____, _____, _____, _____, _____
17. 341, 351, 361, _____, _____, _____, _____, _____, _____
18. 191, 193, _____, _____, 199, _____, _____, _____, _____

Identify rules to describe and create patterns

Write the missing pattern or rule. The first one is done for you.

	Pattern	Rule
1	50 46 42 38 34	Start at 50 and count back by 4s.
2		Start at 73 and count back by 10s.
3	63 66 69 72 75	
4		Start at 86 and count back by 5s.
5	50 55 60 65 70	
6		Start at 60 and count forward by 4s.
7	114 104 94 84 74	
8		Start at 103 and count forward by 5s.
9	132 130 128 126 124	
10		Start at 200 and count back by 3s.

Identify rules to complete number sequences

Work out the pattern for each row of numbers and use it to complete the sequence.

1 2, 6, 10, 14, _____, _____, _____, _____, _____

2 10, 30, 50, _____, _____, _____, _____, _____

3 1, 2, 4, 8, 16, _____, _____, _____, _____, _____

4 640, 320, 160, 80, _____, _____, _____, _____

5 10, 12, 15, 19, _____, _____, _____, _____, _____

6 200, 180, 160, _____, _____, _____, _____, _____

7 5, 6, 8, 9, 11, 12, _____, _____, _____, _____, _____

8 5, 7, 10, 12, 15, 17, _____, _____, _____, _____, _____

9 150, 140, 120, 110, 90, _____, _____, _____, _____

10 10, 15, 25, 30, 40, 45, _____, _____, _____, _____

11 50, 49, 47, 46, 44, 43, _____, _____, _____, _____

12 30, 130, 230, 330, _____, _____, _____, _____, _____

13 750, 700, 650, 600, _____, _____, _____, _____

14 100, 150, 250, 300, 400, _____, _____, _____, _____

15 45, 55, 75, 85, 105, _____, _____, _____, _____, _____

16 220, 210, 190, 180, 160, _____, _____, _____, _____

17 80, 81, 86, 87, 92, _____, _____, _____, _____, _____

18 105, 100, 99, 94, 93, _____, _____, _____, _____

Recognise number patterns

Write the number in each group that does not belong and explain why it is different.

1

50	100	30
20	45	
80	90	60

2

48	43	45
71	42	
47	44	49

3

86	152	70
38	94	
79	126	148

4

71	81	46
31	11	
61	21	51

5

10	25	5
45	30	
17	50	15

6

57	36	23
89	121	
235	93	167

Identify and use rules to create number pairs

Write the rule that connects the numbers on the left to the numbers on the right and use it to complete the missing numbers.

1

3 → 6
5 → 10
8 → 16
10 → ☐
20 → ☐

2

3 → 13
8 → 18
5 → 15
7 → ☐
11 → ☐

3

2 → 10
6 → 30
10 → 50
3 → ☐
5 → ☐

4

24 → 21
19 → 16
8 → 5
30 → ☐
22 → ☐

5

12 → 6
8 → 4
20 → 10
16 → ☐
30 → ☐

6

10 → 15
12 → 17
21 → 26
15 → ☐
24 → ☐

7

9 → 15
16 → 22
4 → 10
12 → ☐
20 → ☐

8

6 → 60
10 → 100
2 → 20
7 → ☐
3 → ☐

9

18 → 13
30 → 25
22 → 17
10 → ☐
36 → ☐

Assessment Patterns

1 Copy and complete these number charts.

a

	154			
163				
173				

b

		95		
			106	
		115		

2 Complete these counting sequences.

a 65, 67, 69, ______, ______, ______, ______, ______, ______, ______

b 254, 264, 274, ______, ______, ______, ______, ______, ______, ______

c 136, 131, 126, ______, ______, ______, ______, ______, ______, ______

3 Write the rule to show how each number pattern has been created.

a 80, 74, 68, 62, 56, 50

b 93, 98, 103, 108, 113, 118

c 117, 127, 137, 147, 157, 167

d 226, 217, 208, 199, 190, 181

e 414, 425, 436, 447, 458, 469

f 162, 142, 122, 102, 82, 62

4 Write the rule that connects the numbers on the left to the numbers on the right and use it to complete the missing numbers.

a
8 → 13
4 → 9
11 → 16
18 → □

b
26 → 16
58 → 48
37 → 27
63 → □

c
4 → 20
7 → 35
3 → 15
8 → □

Important Facts

Times tables

These are the times tables you should know by the end of grade 3.

1 × table	**2 × table**	**5 × table**	**10 × table**
1 × 0 = 0	2 × 0 = 0	5 × 0 = 0	10 × 0 = 0
1 × 1 = 1	2 × 1 = 2	5 × 1 = 5	10 × 1 = 10
1 × 2 = 2	2 × 2 = 4	5 × 2 = 10	10 × 2 = 20
1 × 3 = 3	2 × 3 = 6	5 × 3 = 15	10 × 3 = 30
1 × 4 = 4	2 × 4 = 8	5 × 4 = 20	10 × 4 = 40
1 × 5 = 5	2 × 5 = 10	5 × 5 = 25	10 × 5 = 50
1 × 6 = 6	2 × 6 = 12	5 × 6 = 30	10 × 6 = 60
1 × 7 = 7	2 × 7 = 14	5 × 7 = 35	10 × 7 = 70
1 × 8 = 8	2 × 8 = 16	5 × 8 = 40	10 × 8 = 80
1 × 9 = 9	2 × 9 = 18	5 × 9 = 45	10 × 9 = 90
1 × 10 = 10	2 × 10 = 20	5 × 10 = 50	10 × 10 = 100
1 × 11 = 11	2 × 11 = 22	5 × 11 = 55	10 × 11 = 110
1 × 12 = 12	2 × 12 = 24	5 × 12 = 60	10 × 12 = 120

Ordinal numbers

Ordinal numbers tell the order or position of objects. The first twelve ordinal numbers are:

first	1st
second	2nd
third	3rd
fourth	4th
fifth	5th
sixth	6th
seventh	7th
eighth	8th
ninth	9th
tenth	10th
eleventh	11th
twelfth	12th

Odd and even numbers

An even number is a number that can be divided equally by 2. All even numbers end in 0, 2, 4, 6 or 8.

Examples of even numbers are: 6, 50, 22, 78, 134, 376 and 208.

An odd number is a number that does not divide equally by 2. All odd numbers end in 1, 3, 5, 7 or 9.

Examples of odd numbers are: 5, 43, 61, 97, 219, 507 and 165.

Number pairs for 10, 100 and 1000

0 + 10 = 10	0 + 100 = 100	0 + 1000 = 1000
1 + 9 = 10	10 + 90 = 100	100 + 900 = 1000
2 + 8 = 10	20 + 80 = 100	200 + 800 = 1000
3 + 7 = 10	30 + 70 = 100	300 + 700 = 1000
4 + 6 = 10	40 + 60 = 100	400 + 600 = 1000
5 + 5 = 10	50 + 50 = 100	500 + 500 = 1000
6 + 4 = 10	60 + 40 = 100	600 + 400 = 1000
7 + 3 = 10	70 + 30 = 100	700 + 300 = 1000
8 + 2 = 10	80 + 20 = 100	800 + 200 = 1000
9 + 1 = 10	90 + 10 = 100	900 + 100 = 1000
10 + 0 = 10	100 + 0 = 100	1000 + 0 = 1000

Money

Papua New Guinea uses kina (K) and toea (t). There are 100 toea in 1 kina.

These are pictures of the coins used.

Time

60 minutes = 1 hour	24 hours = 1 day	7 days = 1 week
2 weeks = 1 fortnight	52 weeks = 1 year	12 months = 1 year
365 days = 1 year	366 days = 1 leap year	

Answers

Strand: Number and Application

Pages 2–3

1 a 27 b 45 c 71 d 89 e 36 f 93 g 68 h 54

2 a 62, 56, 40, 28, 18 b 95, 73, 31, 29, 12

3 a thirty-nine b ninety-three c fifty-six d eighty-five e thirty f seventy-eight g twenty-four h sixty-one i forty-seven j eighty-eight k eighteen l eighty-nine

4 a 51 b 38 c 90 d 24 e 43 f 62 g 18 h 95

Pages 3–4

1 a 745 b 917 c 184 d 361 e 536 f 229 g 316 h 970

2 a 850 b 516 c 209 d 731 e 420 f 191

3 a one hundred and fifty-nine b one hundred and five c one hundred and ninety-nine d one hundred and thirty-eight e four hundred and sixty-one f seven hundred and nineteen g three hundred and nine h seven hundred and ninety-nine i three hundred j five hundred and sixty-nine k nine hundred and eighteen l six hundred and forty-four

Pages 4–6

1 a 3rd b Grace c Nonti d Violet e 5th f 9th g Regina h 7th

2 a Ben b 8th c 7th d Kathy e Mea f 4th g Jack h Mary

3 a 6th b 12th c 9th d 8th e 3rd f 5th g 10th h 7th i 4th j 11th

Page 7

1 67 and 25 **2** 84 and 27 **3** 92 and 17 **4** 87 and 30
5 76 and 34 **6** 86 and 16 **7** 95 and 36 **8** 79 and 24

Pages 8–9

1 a 76 b 43 c 95 d 86 e 623 f 508 g 421 h 751 i 216 j 463 k 715 l 806

Teacher to check that other numbers written by students are larger than these ones.

2 a–j. Teacher to check

3 a 411, 505, 570, 604, 722, 885, 937
b 67, 95, 150, 156, 182
c 156, 570, 95, 885, 150, 505. 885 is the largest. 95 is the smallest.

Pages 10–11

1 a 37 b 76 c 50 d 49 e 154 f 321 g 206 h 172 i 237 j 413

2 a 51 b 327 c 14 d 740 e 132 f 308 g 93 h 511 i 485 j 104 k 720 l 49 m 655 n 806 o 217

Page 12

1 Red: 350, 253, 157, 452, 53, 654, 59
Green: 31, 35, 30, 33, 132
Yellow: 268, 478, 518
Blue: 86, 26

2 Teacher to check that all numbers written have 3 tens.

3 Teacher to check.

Page 13

1 72 and 27

2 a 32 b 60 c 83 d 95

3 872, 827, 782, 728, 287, 278

4 a 954 b 459 c 495 d 945

5 a 87 b 217 c 549 d 55 e 238 f 317 g 261 h 533 i 712 j 186 k 354 l 479 m 730 n 905

Page 14

1 20, 10, 15, 5, 35, 25, 30, 40 **2** 80, 100, 95, 85, 75, 110, 90, 105, 115
3 62, 52, 67, 57, 87, 77, 72, 82 **4** 101, 96, 106, 116, 121, 111, 126, 131

Pages 15–16

1 a 60 b 100 c 80 d 140 e 90 f 130

2 a K70 b K120 c K170

3 a 90 b 140 c 60 d 110

4 530 petals

Page 17

220 stars

Page 18

1 64, 69, 74, 79, 84, 89 **2** 53, 63, 73, 83, 93, 103
3 73, 78, 83, 88, 93, 98 **4** 61, 51, 41, 31, 21, 11
5 85, 105, 125, 145, 165, 185 **6** 47, 42, 37, 32, 27, 22
7 160, 140, 120, 100, 80, 60 **8** 117, 127, 137, 147, 157, 167
9 214, 234, 254, 274, 294, 314 **10** 306, 296, 286, 276, 266, 256
11 108, 103, 98, 93, 88, 83 **12** 264, 274, 284, 294, 304, 314
13 489, 469, 449, 429, 409, 389 **14** 202, 207, 212, 217, 222, 227
15 416, 436, 456, 476, 496, 516 **16** 418, 408, 398, 388, 378, 368
17 106, 101, 96, 91, 86, 81 **18** 703, 723, 743, 763, 783, 803

Page 19

1 a 75 b 100 **2** a 90 b 70

3 a 95 b 80 **4** a 105 b 85

Page 20

1 a 120 b 100 **2** a 150 b 140

3 a 130 b 160 **4** a 110 b 90

Page 21

1 14, 48, 72, 90, 6, 34, 24, 66

2 25, 47, 31, 79, 83, 67, 95, 11, 29

3 420, 118, 396, 266, 170, 274, 500, 122, 238, 494

4 511, 383, 407, 113, 695, 829, 523, 799, 185, 907

Page 22

1 Predictions will vary.

2 a odd b odd c even d even e odd f odd
g even h even i odd j even k even l even
m even n even o odd p odd q odd r odd
s odd t even u odd v odd w odd x even

3 Predictions will vary.

4 a even b odd c even d even e odd f even
g odd h odd i even j even k even l odd
m even n even o even p odd q odd r odd
s even t even

5 a–b Teacher to check.

Page 23

1 a 30 b 34 c 29 d 32 e 37 f 35
g 60 h 60 i 62 j 59

2 a 20 b 29 c 20 d 25 e 25 f 22
g 26 h 23 i 25 j 29 k 28 l 21
m 25 n 25

3 a 300 b 220 c 300 d 220

Page 24

1 a 9, 90 b 5, 50 c 9, 90 d 4, 40

2 a 2, 20 b 8, 80 c 6, 60 d 7, 70

3 a 80 b 80 c 100 d 110 e 150 f 110
g 120 h 140

4 a 40 b 30 c 10 d 60 e 30 f 60
g 20 h 40

Page 25

1 57, 67, 77, **87**, 97, 107, 117

2 33, 43, 53, **63**, 73, 83, 93

3 90, 100, 110, **120**, 130, 140, 150

4 121, 131, 141, **151**, 161, 171, 181

5 175, 185, 195, **205**, 215, 225, 235

6 155, 165, 175, **185**, 195, 205, 215

7 62, 72, 82, **92**, 102, 112, 122

8 86, 96, 106, **116**, 126, 136, 146

9 302, 312, 322, **332**, 342, 352, 362

10 254, 264, 274, **284**, 294, 304, 314

11 48, 58, 68, **78**, 88, 98, 108

12 109, 119, 129, **139**, 149, 159, 169

13 430, 440, 450, **460**, 470, 480, 490

Page 26

1 a 4, 5, 40, 50 b 14, 15, 140, 150 c 10, 11, 100, 110
d 16, 17, 160, 170 e 6, 7, 60, 70 f 20, 21, 200, 210

2 a 23 b 31 c 39 d 29 e 25 f 41
g 51 h 61 i 45 j 71 k 101 l 89
m 37 n 35 o 83 p 67 q 49 r 161
s 131 t 55

3 a 112 b 54 c 152 d 70

Page 27

1 a 45 b 66 c 37 d 52 e 74 f 81
g 48 h 32 i 93 j 60 k 88 l 56
m 164 n 137 o 241 p 328 q 275 r 207
s 181 t 374 u 296 v 343 w 112 x 258

2 a 43 b 62 c 29 d 84 e 77 f 51
g 66 h 38 i 15 j 54 k 86 l 47
m 119 n 94 o 155 p 279 q 327 r 191
s 148 t 206 u 312 v 113 w 265 x 331

3 a 90, 99, 108, 117, 126, 135, 144, 153
b 150, 141, 132, 123, 114, 105, 96, 87
c 242, 251, 260, 269, 278, 287, 296, 305

Page 28

1 a 36 b 62 c 45 d 41 e 58 f 57
g 27 h 78 i 52 j 82 k 66 l 59
m 118 n 143 o 242 p 272 q 140 r 189
s 189 t 331 u 227 v 296 w 306 x 198

2 a 91, 102, 113, 124, 135, 146, 157, 168
b 91, 80, 69, 58, 47, 36, 25, 14
c 309, 320, 331, 342, 353, 364, 375, 386
d 180, 169, 158, 147, 136, 125, 114, 103

Page 29

1 a

+	9	10	11
32	41	42	43
67	76	77	78
188	197	198	199
145	154	155	156
209	218	219	220

b

–	9	10	11
84	75	74	73
46	37	36	35
117	108	107	106
231	222	221	220
183	174	173	172

2 a 64, 55, 65, 54, 63 b 8, 18, 29, 20, 31
c 99, 90, 79, 88, 77 d 147, 157, 148, 139, 149
e 204, 195, 184, 194, 185 f 184, 193, 204, 195, 205
g 250, 261, 251, 240, 249 h 318, 307, 316, 306, 295

Page 30

1 a 175 b 158 c 164 d 181 e 133 f 127
g 146 h 287 i 324 j 255 k 574 l 483
m 626 n 413

2 a 65 b 182 c 289 d 118 e 29 f 363
g 254 h 430 i 337 j 236 k 52 l 631
m 405 n 532

Page 31

1 +2, –1, +4, –5 **2** +2, –4, –2, +3 **3** –4, –1, +5, +10

4 –3, +2, +3, +5 **5** –6, –2, +5, –2 **6** +3, +2, –7, +4

7 +6, –5, –2, +4 **8** +4, +6, –3, +6 **9** –3, +7, +4, –6

10 –3, +8, +6, –7 **11** –6, +7, +4, –5 **12** –3, +7, +3, –4

Page 32

1 70, 10, 40, 20, 30, 50, 60, 90

2 a 40 b 70 c 10 d 60 e 50 f 80
g 35 h 42 i 27 j 49 k 54 l 13
m 61 n 22 o 76 p 68 q 57 r 75
s 82 t 31 u 73 v 29 w 44 x 62

3 200 and 800; 700 and 300; 950 and 50; 840 and 160; 510 and 490; 750 and 250

4 a 600 b 850 c 650 d 720 e 810 f 540
g 120 h 430 i 390 j 70 k 580 l 240
m 195 n 405 o 245 p 715 q 85 r 625

Page 33

1 a 9 b 7 c 9 d 7 e 8 f 10
g 9 h 7 i 6 j 4 k 3 l 9
m 6 n 8 o 11 p 8 q 6 r 9
s 6 t 9 u 8 v 11 w 6 x 10

2 a 9 b 20 c 16 d 6 e 5 f 5
g 12 h 14 i 8 j 20 k 18 l 6
m 6 n 17 o 6 p 18 q 17 r 6
s 7 t 11 u 6 v 18 w 16 x 9

Pages 34–35

1 a yes b no c no d yes e no f yes
g no h yes i yes j no k yes l yes
m no n no o yes p yes q yes r no
s yes t yes u yes v no w yes x no
y no z no

2 a yes b yes c no d no e yes f yes
g yes h no i yes j no k yes l yes
m yes n no o yes p no q yes r yes
s yes t yes u no v yes w yes x no
y yes z no

Page 36

Game

Page 37

Game

Page 38

1 a 8 b 16 c 12 d 10 e 22 f 18
g 6 h 14 i 20 j 2 k 24 l 4

2 a 10 b 3 c 5 d 8 e 4 f 7
g 1 h 12 i 2 j 6 k 11 l 9

3 a 9 b 3 c 6 d 2 e 8 f 5
g 4 h 10 i 7

4 a 10 b 16 c 12 d 14 e 22 f 18
g 24 h 8

Page 39

1 a

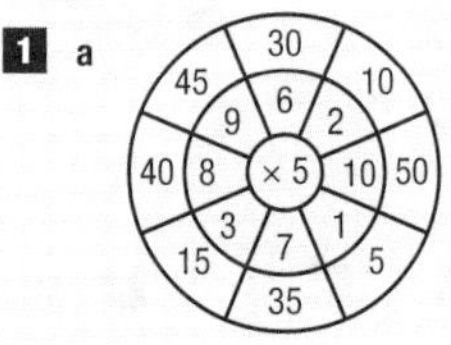

b

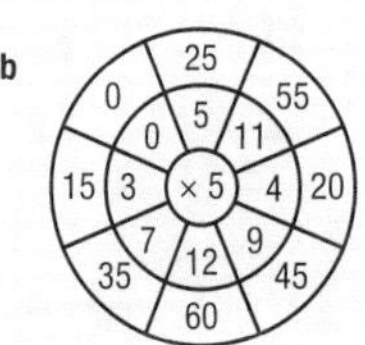

2 a 5 b 50 c 3 d 8 e 6 f 1
g 55 h 9 i 20 j 7 k 12 l 2

3 a 5 × 4 = 20 b 5 × 9 = 45 c 5 × 2 = 10 d 5 × 6 = 30
e 5 × 3 = 15 f 5 × 5 = 25 g 5 × 8 = 40 h 5 × 12 = 60
i 5 × 7 = 35

4 a 5 b 8 c 3 d 10 e 7 f 4

Page 40

1 50, 80, 20, 60, 0, 90, 10, 30, 70, 100, 40

2 a 9 b 4 c 1 d 7 e 2 f 5
g 10 h 6 i 3

3 a 10 × 10 = 100 b 10 × 6 = 60 c 10 × 8 = 80 d 10 × 0 = 0
e 10 × 3 = 30 f 10 × 12 = 120 g 10 × 5 = 50 h 10 × 7 = 70
i 10 × 2 = 20 j 10 × 9 = 90 k 10 × 4 = 40 l 10 × 11 = 110

4 a

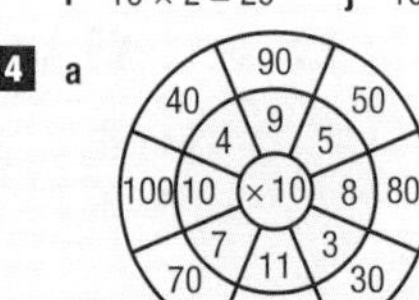

b

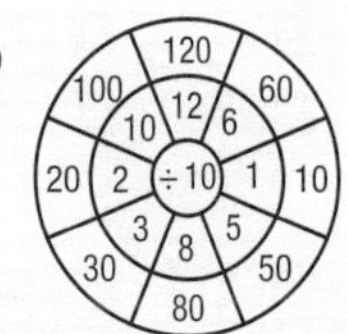

Page 41

1

×	4	1	9	7	0	6	2	8	3	10	5
1	4	1	9	7	0	6	2	8	3	10	5
2	8	2	18	14	0	12	4	16	6	20	10
5	20	5	45	35	0	30	10	40	15	50	25
10	40	10	90	70	0	60	20	80	30	100	50

2 **a** 18 **b** 6 **c** 6 **d** 8 **e** 10 **f** 8 **g** 25 **h** 5 **i** 40 **j** 7 **k** 1 **l** 45 **m** 2 **n** 5 **o** 7 **p** 14 **q** 8 **r** 11 **s** 1 **t** 5 **u** 110

3 **a** 2 × 15, 5 × 6, 10 × 3 **b** 5 × 3 **c** 2 × 12 **d** 2 × 10, 5 × 4, 10 × 2 **e** 5 × 9 **f** 2 × 9 **g** 5 × 5 **h** 2 × 30, 5 × 12, 10 × 6 **i** 2 × 7

Pages 42–43

1 **a** 87 **b** 96 **c** 89 **d** 79 **e** 92 **f** 93 **g** 104 **h** 113 **i** 132 **j** 149 **k** 111 **l** 181

2 **a** 131 **b** 132 **c** 131 **d** 144 **e** 137 **f** 143 **g** 115 **h** 142 **i** 124 **j** 141 **k** 160 **l** 122

Page 44

1 488 **2** 479 **3** 668 **4** 791 **5** 840 **6** 1031 **7** 824 **8** 663 **9** 722

Page 45

1 125 **2** 181 **3** 142 **4** 177 **5** 96 **6** 181 **7** 222 **8** 222 **9** 451 **10** 429 **11** 452 **12** 181

The code is VERY WELL DONE

Page 46

1 134 bags **2** 202 people **3** K221 **4** 223 kilograms **5** 625 visitors **6** 264 trees **7** 523 spectators

Pages 47–48

1 **a** 42 **b** 42 **c** 62 **d** 40 **e** 58 **f** 16 **g** 29 **h** 69 **i** 32 **j** 45 **k** 35

2 **a** 41 **b** 49 **c** 53 **d** 46 **e** 33 **f** 34 **g** 57 **h** 48 **i** 37 **j** 37 **k** 67 **l** 46

Page 49

1 262 **2** 343 **3** 642 **4** 294 **5** 659 **6** 538 **7** 372 **8** 591 **9** 757

Page 50

1 115 **2** 431 **3** 171 **4** 267 **5** 484 **6** 648 **7** 277 **8** 277 **9** 564

Page 51

1 K355 **2** 79 people **3** 79 trees **4** 178 kilometres **5** K149 **6** 125 people **7** 255 kilograms

Page 52

1 13 + 13 + 13 = 39 and 3 × 13 = 39

2 22 + 22 + 22 + 22 = 88 and 4 × 22 = 88

3 18 + 18 = 36 and 2 × 18 = 36

4 21 + 21 + 21 + 21 + 21 + 21 = 126 and 6 × 21 = 126

5 12 + 12 + 12 + 12 = 48 and 4 × 12 = 48

6 11 + 11 + 11 + 11 + 11 = 55 and 5 × 11 = 55

7 11 + 11 + 11 + 11 = 44 and 4 × 11 = 44

8 14 + 14 = 28 and 2 × 14 = 28

9 12 + 12 + 12 = 36 and 3 × 12 = 36

Page 53

1 **a** 84 **b** 96 **c** 48 **d** 39 **e** 84 **f** 99 **g** 147 **h** 155

2 **a** 78 **b** 90 **c** 96 **d** 172 **e** 225 **f** 256 **g** 165 **h** 185 **i** 168 **j** 315 **k** 136 **l** 318

Page 54

1 27 × 6 = 162 **2** 24 × 5 = 120 **3** 45 × 5 = 225 **4** 48 × 4 = 192 **5** 36 × 4 = 144 **6** 56 × 3 = 168 **7** 32 × 8 = 256 **8** 18 × 7 = 126

(Students should set out all multiplications vertically.)

Page 55

1 **a** 16 ÷ 8 = 2 **b** 35 ÷ 5 = 7 **c** 30 ÷ 6 = 5 **d** 24 ÷ 3 = 8 **e** 36 ÷ 9 = 4 **f** 80 ÷ 10 = 8 **g** 16 ÷ 4 = 4 **h** 18 ÷ 6 = 3

2 **a** 32 ÷ 4 = 8 and 32 ÷ 8 = 4 **b** 63 ÷ 9 = 7 and 63 ÷ 7 = 9 **c** 12 ÷ 4 = 3 and 12 ÷ 3 = 4 **d** 40 ÷ 5 = 8 and 40 ÷ 8 = 5 **e** 21 ÷ 3 = 7 and 21 ÷ 7 = 3 **f** 27 ÷ 3 = 9 and 27 ÷ 9 = 3 **g** 22 ÷ 11 = 2 and 22 ÷ 2 = 11 **h** 15 ÷ 3 = 5 and 15 ÷ 5 = 3 **i** 14 ÷ 7 = 2 and 14 ÷ 2 = 7 **j** 20 ÷ 4 = 5 and 20 ÷ 5 = 4

Page 56

1 **a** 16 ÷ 4 = 4 **b** 12 ÷ 3 = 4 **c** 15 ÷ 3 = 5 **d** 20 ÷ 2 = 10 **e** 18 ÷ 3 = 6 **f** 12 ÷ 4 = 3 **g** 21 ÷ 3 = 7 **h** 14 ÷ 2 = 7 **i** 25 ÷ 5 = 5

2 a $48 \div 8 = 48 - 8 - 8 - 8 - 8 - 8 - 8 = 0$, so $48 \div 8 = 6$
b $35 \div 5 = 35 - 5 - 5 - 5 - 5 - 5 - 5 - 5 = 0$, so $35 \div 5 = 7$
c $50 \div 10 = 50 - 10 - 10 - 10 - 10 - 10 = 0$, so $50 \div 10 = 5$
d $36 \div 6 = 36 - 6 - 6 - 6 - 6 - 6 - 6 = 0$, so $36 \div 6 = 6$
e $18 \div 3 = 18 - 3 - 3 - 3 - 3 - 3 - 3 = 0$, so $18 \div 3 = 6$
f $16 \div 4 = 16 - 4 - 4 - 4 - 4 = 0$, so $16 \div 4 = 4$
g $20 \div 5 = 20 - 5 - 5 - 5 - 5 = 0$, so $20 \div 5 = 4$
h $28 \div 7 = 28 - 7 - 7 - 7 - 7 = 0$, so $28 \div 7 = 4$
i $30 \div 5 = 30 - 5 - 5 - 5 - 5 - 5 - 5 = 0$, so $30 \div 5 = 6$
j $40 \div 10 = 40 - 10 - 10 - 10 - 10 = 0$, so $40 \div 10 = 4$
k $15 \div 3 = 15 - 3 - 3 - 3 - 3 - 3 = 0$, so $15 \div 3 = 5$
l $25 \div 5 = 25 - 5 - 5 - 5 - 5 - 5 = 0$, so $25 \div 5 = 5$

Page 57

1 a 2 rem. 2 b 2 c 2 rem. 3 d 6 rem. 3
e 10 rem. 1 f 0 g 6 rem. 2 h 6 rem. 3
i 9 j 8 rem. 2 k 4 l 7 rem. 1
m 4 rem. 2 n 8 rem. 1 o 6 p 6 rem. 1
q 4 rem. 3 r 4 rem. 4 s 3 rem. 1 t 7
u 11 v 5 rem. 1 w 9 rem. 1 x 5

2 a 2 b 10 c 6 d 90 e 18 f 40
g 5 h 6 i 6 j 4 k 4 l 6
m 10 n 4 o 5 p 6 q 7 r 7

Page 58

Game

Page 59

1 6 dresses **2** 184 books **3** 138 pieces **4** 8 balls
5 8 packages **6** 235 kilograms **7** 11 people

Page 60

1 88 have been sold **2** 11 kilometres **3** 323 people
4 K69 **5** 464 litres **6** 8 bags
7 95 kilograms

Page 61

1 yes **2** no **3** yes **4** no **5** yes **6** yes
7 yes **8** no **9** no **10** yes **11** yes **12** yes
13 no **14** no **15** yes

Page 62

1 yes **2** yes **3** no **4** no **5** yes **6** no
7 no **8** yes **9** yes **10** no **11** yes **12** yes
13 yes **14** no **15** no

Page 63

1 yes **2** yes **3** no **4** no **5** yes **6** yes
7 no **8** no **9** yes **10** yes **11** yes **12** no
13 yes **14** no **15** no

Page 64

1 $\frac{1}{2}$, one half: D and F **2** $\frac{1}{4}$, one quarter: A and H **3** $\frac{1}{3}$, one third: G and J
4 $\frac{1}{5}$, one fifth: C and E **5** $\frac{1}{10}$, one tenth: B and I

Page 65

1 $\frac{1}{2}$: H, J, K and L **2** $\frac{1}{4}$: F, G, N and O **3** $\frac{1}{3}$: A, D, P and Q
4 $\frac{1}{5}$: B, C and I **5** $\frac{1}{10}$: E and M

Page 66

1 $\frac{1}{3}$, one third **2** $\frac{1}{4}$, one quarter **3** $\frac{1}{5}$, one fifth
4 $\frac{1}{3}$, one third **5** $\frac{1}{10}$, one tenth **6** $\frac{1}{2}$, one half
7 $\frac{1}{10}$, one tenth **8** $\frac{1}{3}$, one third **9** $\frac{1}{2}$, one half
10 $\frac{1}{5}$, one fifth **11** $\frac{1}{4}$, one quarter **12** $\frac{1}{2}$, one half

Page 67

1 b **2** c **3** b **4** a **5** a

Pages 68–69

The following answers use the least number of notes and coins. There are other possibilities.

1 a K5, K2, K1 and 50t
b K5, K1, 50t, 20t and 5t
c K10, K5, 50t and 20t
d K20, K5, K2, K1, 20t, 10t and 5t
e K10, K5, K2, K1, 50t, 20t and 20t

The following answers use the least number of notes and coins. There are other possibilities.

2 a K20, K10, K5, K1 and 50t
b K20, K5, K2, K2, 20t, 20t and 5t
c K50, K5, K2, 50t and 10t
d K10, K5, K2, K1, 20t and 5t
e K20, K20, K2, K1, 50t, 20t and 5t

Page 70

The following answers use the least number of notes and coins. There are other possibilities.

1 K1, 50t and 5t = K1.55 **2** 50t and 5t = 55t
3 K1 and 20t = K1.20 **4** K5 and 20t = K5.20
5 K1, 20t and 5t = K1.25 **6** K10, 20t and 10t = K10.30
7 K2, K2 and 50t = K4.50 **8** K10 and K1 = K11
9 K10, K2, K2 and 20t = K14.20 **10** K5, K1 and 20t = K6.20
11 K20, K2 and 50t = K22.50 **12** K10, K2 and 20t = K12.20

Pages 71–73

1 a K1.30 b K1.50 c 75t d K1.15

2 a K90 b K130 c K82 d K85

3 a K1.30 b K1.40 c K1.60 d 95t e K1.90 f K1.45
g K1.75 h K1.05 i 90t j K1.55 k K2.70 l K2.75

4 Box l has the most money (K2.75).

5 Box i has the least money (90t).

6 The coins on the page total K4.40.

Page 74

The following coins should be drawn.

1 50t and 10t **2** 20t, 10t and 5t **3** 20t, 20t and 5t

4 20t and 10t or 20t, 5t and 5t

5 50t, 20t and 5t or 20t, 20t, 20t, 10t and 5t

6 20t, 20t, 10t and 5t or 20t, 10t, 10t, 10t and 5t

Page 75

The following coins should be drawn. These answers use the least number of coins.

1 K1 and 50t **2** 50t, 20t and 10t

3 50t, 20t and 20t **4** 50t, 50t, 20t, 10t and 5t

5 K1, 50t and 5t **6** 20t, 20t, 20t, 20t, 20t and 10t

Page 76

1 a K3.10 b K3.25 c K6.25 d K3.20 e K5.10
f K9.15 g K10.05 h K10.05 i K9

2 a K2.50 b K2.55 c K1.65 d K4.55 e K1.80
f K3.65 g K2.25 h K3.80 i K2.85

3 a K5 b K2.40 c K3.20 d K17.20 e K16
f K6.45 g K27 h K22.50

4 a K2.10 b K4.30 c K3 d K1.20 e K5.40
f K2.50 g K2.50 h K1.30

Page 77

1 K8.35 × 3 = K25.05 **2** K20.00 – K12.85 = K7.15

3 K9.50 + K2.65 = K12.15 **4** K35.50 + K31.70 = K67.20

5 K9.50 × 5 = K47.50 **6** K35.50 ÷ 5 = K7.10

7 K12.85 – K8.35 = K4.50 **8** K30.40 ÷ 4 = K7.60

Assessment

Pages 78–79

1 b **2** c **3** b **4** d **5** c **6** c

7 b **8** c **9** b **10** a **11** d **12** a

13 b **14** d **15** c **16** a **17** b

Assessment

Pages 80–83

1 a 24 b 27 c 28 d 29

2 a 90 b 50 c 30 d 110 e 50 f 80
g 130 h 50

3 a 64 b 106 c 133 d 137 e 202 f 115
g 261 h 359

4 a 49 b 35 c 83 d 31 e 73 f 43
g 109 h 55

5 a 37 b 47 c 45 d 74 e 138 f 141
g 218 h 234

6 a 154 b 273 c 316 d 62

7 a 55 b 27 c 62 d 39 e 46 f 73

8 a 400 b 650 c 730 d 590 e 220 f 305

9 a 9 b 8 c 9 d 7 e 18 f 8
g 12 h 9

10 a 16 b 24 c 30 d 45 e 30 f 80
g 55 h 15 i 60 j 120 k 14 l 18

11 a 6 b 4 c 4 d 8 e 10 f 3
g 7 h 12 i 6 j 12 k 5 l 7

12 a 89 b 117 c 124 d 122 e 577 f 841
g 831 h 441

13 a 43 b 33 c 49 d 36 e 121 f 151
g 245 h 663

14 a 126 b 124 c 185 d 496

15 a 4 rem. 3 b 3 rem. 1 c 7 rem. 2 d 9 rem. 1

16 a 167 kilometres b K232 c K12
d 168 litres e 49 seedlings

Assessment

Page 84

1 a yes b no c no d yes

2 a yes b yes c no d no

3 a no b yes c yes d yes

4 a $\frac{1}{5}$ b $\frac{1}{3}$ c $\frac{1}{2}$ d $\frac{1}{10}$ e $\frac{1}{4}$ f $\frac{1}{10}$
g $\frac{1}{5}$ h $\frac{1}{4}$

Assessment

Page 85

1 a–c Teacher to check.

2 a K10, K2, 50t and 10t b K20, K5, K2, K1, 20t, 20t and 5t c K10, K5, K2, 50t, 20t and 10t

3 a K10 and 20t b K20, K2, K2 and 5t c K20, K1, 20t, 10t and 5t

4 a K2.60 b K3.55 c K3.30

5 a K2.50 b K2.50 c K6.15 d K3.40 e K1.05 f K2.85

Strand: Measurement

Page 86

1 Line A = 7 cm; Line B = 5 cm; Line C = 10 cm; Line D = 12 cm; Line E = 13 cm; Line F = 15 cm; Line G = 9 cm; Line H = 14 cm

2 Line F **3** 5 cm

4 a 5 cm b 4 cm c 1 cm

Pages 87–88

1 a 20 cm b 20 cm c 32 cm d 26 cm e 24 cm

2 c has the longest perimeter

3 Guesses will vary. **4** Guesses will vary.

5 a 20 cm b 34 cm c 29 cm d 22 cm

Page 89

1 7 square units **2** 8 square units **3** 10 square units

4 6 square units **5** 8 square units

Page 90

1 a = 9; b = 7; c = 8; d = 5; e = 4

2 a **3** e

4 a = 4; b = 4; c = 3 ; d = 3

Page 91

1 a = 9; b = 7; c = 10; d = 10; e = 14; f = 6

2 e **3** f **4** c and d

Page 92

1 a–j Teacher to check.

2 a heavier b lightest c heaviest d heavier

Page 93

A = half-past 2, two-thirty and 2:30
B = seven o'clock and 7:00
C = four o'clock and 4:00
D = half-past 11, eleven-thirty and 11:30
E = ten o'clock and 10:00
F = half-past 6, six-thirty and 6:30

Page 94

1 10:30 **2** 4:00 **3** 12:00 **4** 3:30 **5** 9:00

6 8:30 **7** 5:30 **8** 7:00

Page 95

1 7 o'clock, 7:00 **2** 1 o'clock, 1:00 **3** 4 o'clock, 4:00

4 12 o'clock, 12:00 **5** 6 o'clock, 6:00 **6** 2 o'clock, 2:00

7 3 o'clock, 3:00 **8** 10 o'clock, 10:00 **9** 8 o'clock, 8:00

10 11 o'clock, 11:00 **11** 9 o'clock, 9:00 **12** 5 o'clock, 5:00

Page 96

1 half-past 5, 5:30 **2** half-past 2, 2:30 **3** half-past 9, 9:30

4 half-past 7, 7:30 **5** half-past 10, 10:30 **6** half-past 4, 4:30

7 half-past 1, 1:30 **8** half-past 8, 8:30 **9** half-past 11, 11:30

10 half-past 6, 6:30 **11** half-past 12, 12:30 **12** half-past 3, 3:30

Page 97

1 a 1 week b 15 days c 1 hour d 75 minutes
e 18 months f 2 years g 9 days h 2 hours
i 1 year j 19 days k 37 months l 150 minutes
m 3 weeks n 50 days o 3 hours p 10 years
q 6 weeks r 40 days s 1 year t 70 weeks
u 2 months v 30 hours

2 a 200 hours, 10 days, 2 weeks, 1 month
b 12 hours, 1000 minutes, 20 hours, 1 day
c 40 weeks, 1 year, 16 months, 2 years
d 100 days, 2 years, 30 months, 3 years
e 12 weeks, 6 months, 300 days, 1 year

Page 98

1 4 **2** 5 **3** 4

4 January, March, May, July, August, October, December

5 April, June, September, November

6 28 **7** Thursday **8** Wednesday **9** Monday

10 Tuesday **11** Friday **12** Friday **13** 18th

14 8th **15** 23rd **16** 120 **17** 122

Assessment

Pages 99–101

1 a 6 cm b 8 cm c 11 cm d 7 cm

2 a 12 cm b 14 cm c 16 cm

3 a B b 4 square units c 21 square units

4 a B b 2 cubes c 17 cubes

5 a 6:00 b half past nine, 9:30 c half past two, $\frac{1}{2}$ past 2
d $\frac{1}{2}$ past 7, 7:30 e eleven o'clock, 11 o'clock

6 a 8 o'clock, 8:00 b $\frac{1}{2}$ past 3, 3:30 c 2 o'clock, 2:00

7 a

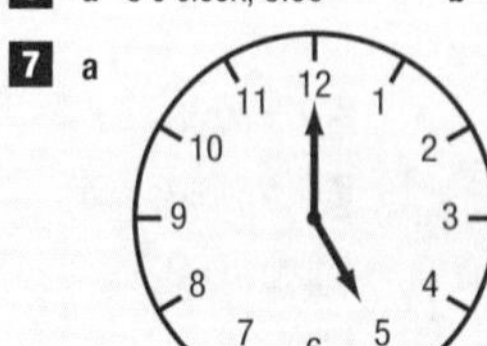

b

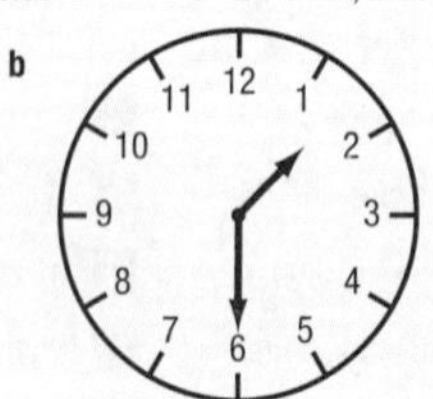

c

d

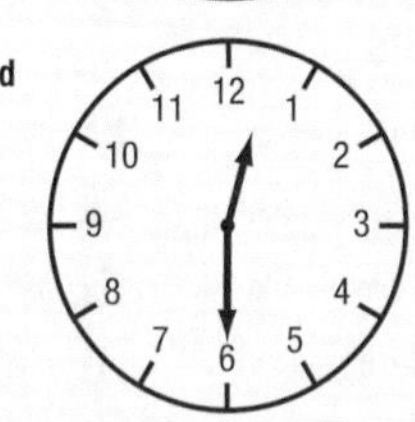

8 a 9 days b 1 hour c 2 years d 6 weeks
e 2 weeks f 1 day g 450 days h 15 months
i 2 hours j 150 weeks k 2 months l 100 hours

Strand: Space and Shape

Page 102

square = F and K; circle = A and J; triangle = B, H and M; rectangle = D and G; hexagon = E and I; trapezium = C and L

Page 103

cube = C and K; sphere = B and F; cylinder = E, G and I; pyramid = D, H and L; cone = A and J

Page 104

1 Shapes A, D, G, J, K and P **2** Shapes B, I and N

3 Shapes G, J and P **4** Shapes C, E, L and O

5 Shapes F, H and M

Page 105

1 Shapes C, F and H **2** Shapes C, F and H

3 Shapes A, G and L **4** Shapes B, D, E, I and K

5 Shapes A, G, J and L **6** Shapes A, G and L

Page 106

1 6 **2** 6 **3** 3 **4** 12 **5** 10 **6** 3

7 4 **8** 6 **9** 1

Page 107

1 north **2** south **3** east **4** west

5 west **6** south **7** north **8** south

9 south, north **10** west, east **11** east, south **12** north, west

Page 108

1 b, d, e and g are right angles **2** c, f and h are straight angles

Assessment

Pages 109–111

1 c is not a circle **2** b is not a square

3 a is not a cylinder **4** c is not a pyramid

5 b, c and e have four corners **6** a, b and e have six faces

7 a 3 b 9 c 4

8 a east b north c south d north e west

9 a, b, d and e are right angles **10** b, c and e are straight angles

Strand: Chance and Data

Page 112

1 impossible **2** impossible **3** certain **4** impossible

5 possible **6** possible

Page 113

Game

Page 114

1 a 11 b 7 c 71 d 1 e 9 f 15
g 3

2 a 10 b 12 c bananas d 13 e 3

Pages 115–116

1 a 6 b 5 c Friday d Wednesday
e 9 f 5 g 2
h Tuesday, Wednesday, Friday and Sunday
i Monday and Thursday, Tuesday and Sunday
j 3 k 37 eggs l 9

2 a Elsie b Nonti c 6 d 17 e 32 f 5
g 5 h 13 i 7 j Grace and Inala

Page 117

1 8 **2** 2 **3** 1 **4** green and pink

5 13 **6** 5 **7** 3 **8** 29

9 blue, yellow, red, orange, green and pink **10** 9

Page 118

1–4 Teacher to check.

Assessment

Page 119

1 a 7 b 16 c K1 d K1.30 e 3

2 a 5 b 4 c 30 d 2 and 6 e 2 f 6

Strand: Patterns

Page 120

1

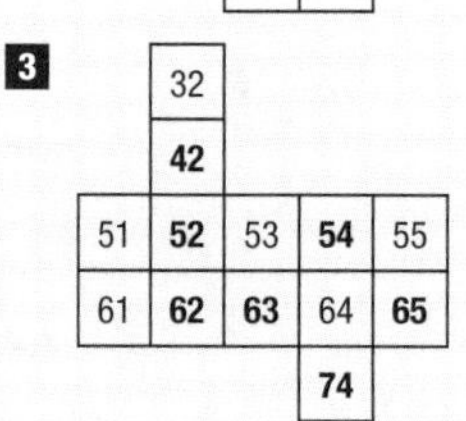

21	22	23	
31	32	33	
41	42	**43**	**44**
		53	**54**
		63	**64**
		73	**74**

2

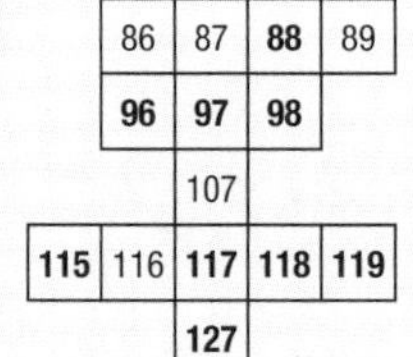

33	**34**				
43	44				
53	**54**	55	**56**	**57**	58
63	64	**65**	**66**	**67**	68

3

	32			
	42			
51	**52**	53	**54**	55
61	**62**	**63**	64	**65**
			74	

4

	86	87	**88**	89
	96	**97**	**98**	
		107		
115	116	**117**	**118**	**119**
		127		

5

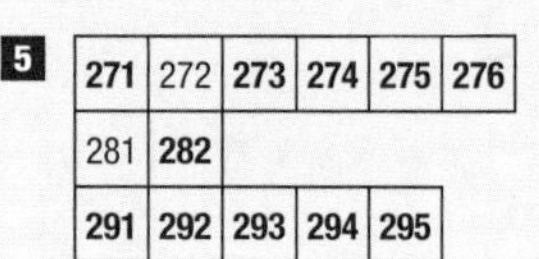

271	272	**273**	**274**	**275**	**276**
281	**282**				
291	**292**	**293**	**294**	**295**	

6

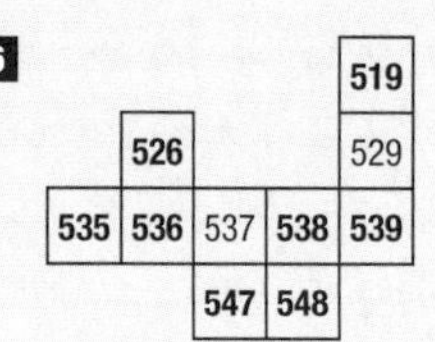

				519
	526			529
535	**536**	537	**538**	**539**
		547	**548**	

Page 121

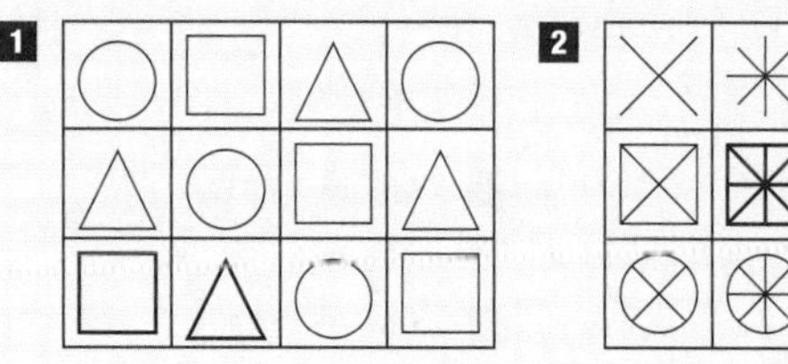

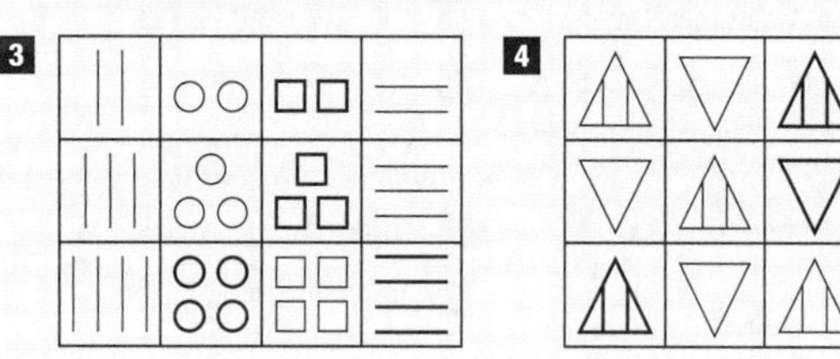

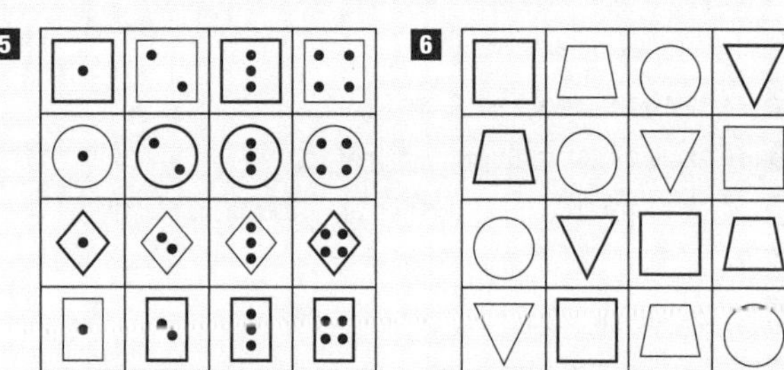

Page 122

1 35, 34, 33, 32, 31, 30, 29, 28, 27, 26

2 50, 52, 54, 56, 58, 60, 62, 64, 66, 68

3 44, 49, 54, 59, 64, 69, 74, 79, 84, 89

4 19, 21, 23, 25, 27, 29, 31, 33, 35, 37

5 23, 33, 43, 53, 63, 73, 83, 93, 103, 113, 123

6 58, 63, 68, 73, 78, 83, 88, 93, 98, 103

7 67, 65, 63, 61, 59, 57, 55, 53, 51, 49

8 151, 141, 131, 121, 111, 101, 91, 81, 71, 61

9 73, 75, 77, 79, 81, 83, 85, 87, 89, 91

10 92, 87, 82, 77, 72, 67, 62, 57, 52, 47

11 100, 102, 104, 106, 108, 110, 112, 114, 116

12 160, 155, 150, 145, 140, 135, 130, 125, 120

13 200, 210, 220, 230, 240, 250, 260, 270, 280

14 235, 225, 215, 205, 195, 185, 175, 165, 155

15 187, 185, 183, 181, 179, 177, 175, 173, 171

16 284, 289, 294, 299, 304, 309, 314, 319, 324

17 341, 351, 361, 371, 381, 391, 401, 411, 421

18 191, 193, 195, 197, 199, 201, 203, 205, 207

Page 123

1 Done

2 73, 63, 53, 43, 33

3 Start at 63 and count forward by 3s.

4 86, 81, 76, 71, 66

5 Start at 50 and count forward by 5s.

6 60, 64, 68, 72, 76

7 Start at 114 and count back by 10s.

8 103, 108, 113, 118, 123

9 Start at 132 and count back by 2s.

10 200, 197, 194, 191, 188

Page 124

1 18, 22, 26, 30, 34 (add 4)

2 70, 90, 110, 130, 150 (add 20)

3 32, 64, 128, 256, 512 (double)

4 40, 20, 10, 5 (halve)

5 24, 30, 37, 45, 54 (add 5, add 6, add 7 and so on)

6 140, 120, 100, 80, 60 (subtract 20)

7 14, 15, 17, 18, 20 (add 1, add 2, add 1, add 2 …)

8 20, 22, 25, 27, 30 (add 2, add 3, add 2, add 3 …)

9 80, 60, 50, 30, 20 (subtract 10, subtract 20, subtract 10, subtract 20 …)

10 55, 60, 70, 75 (add 10, add 5, add 10, add 5 …)

11 41, 40, 38, 37 (subtract 1, subtract 2, subtract 1, subtract 2 …)

12 430, 530, 630, 730, 830 (add 100)

13 550, 500, 450, 400 (subtract 50)

14 450, 550, 600, 700 (add 50, add 100, add 50, add 100 …)

15 115, 135, 145, 165, 175 (add 10, add 20, add 10, add 20 …)

16 150, 130, 120, 100 (subtract 10, subtract 20, subtract 10, subtract 20 …)

17 93, 98, 99, 104, 105 (add 1, add 5, add 1, add 5 …)

18 88, 87, 82, 81 (subtract 5, subtract 1, subtract 5, subtract 1 …)

Page 125

1 45; other numbers have 0 units

2 71; other numbers have 4 tens

3 79; other numbers are even

4 46; other numbers have 1 unit

5 17; other numbers are multiples of 5

6 36; other numbers are odd

Page 126

1 double the number on the left; 20, 40

2 add 10 to the number on the left; 17, 21

3 multiply the number on the left by 5; 15, 25

4 subtract 3 from the number on the left; 27, 19

5 halve the number on the left; 8, 15

6 add 5 to the number on the left; 21, 29

7 add 6 to the number on the left; 18, 26

8 multiply the number on the left by 10; 70, 30

9 subtract 5 from the number on the left; 5, 31

Assessment

Page 127

1 a

153	154	**155**		
163	**164**	**165**	**166**	**167**
173	**174**			
	184	**185**		

b

		95	**96**	
103	**104**	**105**	106	**107**
113	**114**	115	**116**	
		125	**126**	**127**

2 a 71, 73, 75, 77, 79, 81, 83
b 284, 294, 304, 314, 324, 334, 344
c 121, 116, 111, 106, 101, 96, 91

3 a subtract 6
b add 5
c add 10
d subtract 9
e add 11
f subtract 20

4 a add 5 to the number on the left; 23
b subtract 10 from the number on the left; 53
c multiply the number on the left by 5; 40